HEALTH AND THE NEW MEDIA
Technologies Transforming
Personal and Public Health

HEALTH AND THE NEW MEDIA
Technologies Transforming
Personal and Public Health

Edited by

Linda M. Harris
Center for Health Policy Research,
George Washington University

Routledge
Taylor & Francis Group

NEW YORK AND LONDON

First Published by

Lawrence Erlbaum Associates, Inc., Publishers
10 Industrial Avenue
Mahwah, New Jersey 07430

Transferred to Digital Printing 2009 by Routledge
270 Madison Ave, New York NY 10016
2 Park Square, Milton Park, Abingdon, Oxon, OX14 4RN

Cover design by Mairav Salomon-Dekel

Library of Congress Cataloging-in-Publication Data
Health and the new media : technologies transforming personal and
    public health / Linda M. Harris, editor.
            p.   cm.
        "This book sprang from a Forum on Interactive Multimedia and
    Health Care, held November 1992 in Washington, DC"—Pref.
        Includes bibliographical references and index.
        ISBN 0-8058-1569-4 (alk. paper).—ISBN 0-8058-1954-1 (pbk.: alk. paper)
        1. Communication in medicine—Congresses. 2. Interactive
    multimedia—Congresses. 3. Medical informatics—Congresses.
    4. Internet (Computer network)—Congresses. 5. Telecommunication in medicine—
    Congresses. 6. Health education—Congresses. 7. Public health—Congresses. I. Harris,
    Linda M.
    R118.H43   1995
    610'.285—dc20   95-7208
                                                          CIP

Publisher's Note
The publisher has gone to great lengths to ensure the quality of this
reprint but points out that some imperfections in the original may be apparent.

Contents

List of Contributors

Sandra Cheiten, Director, Multimedia Education, ABC News InterActive

Chris Dede, PhD, Director, Center for Interactive Education Technology, George Mason University

Mary Jo Deering, PhD, Staff Director, Health Communication, Office of Disease Prevention and Health Promotion, U.S. Public Health Service

Joseph Eaglin, MACRO International

Francis Dummer Fisher, Visiting Scholar, LBJ School of Public Affairs, University of Texas at Austin

Lynn Fontana, Center for Interactive Education Technology, George Mason University

G. Anthony Gorry, PhD, Vice President for Research and Information Technology, Rice University

Linda M. Harris, PhD, Senior Research Scientist, Center for Health Policy Research, George Washington University

Joseph V. Henderson, MD, Director, Interactive Media Laboratory, Dartmouth Medical School

C. Everett Koop, MD, The C. Everett Koop Institute, Dartmouth University

Julia A. Marsh, Associate, Technology Futures, Inc.

Michael D. McDonald, Dr.PH, Health and Telecommunication Office, The C. Everett Koop Foundation, Washington, DC

J. Michael McGinnis, MD, Deputy Assistant Secretary for Health (Health Promotion and Disease Prevention), U.S. Public Health Service

Kevin Patrick, MD, Senior Advisor for Health Communication Technology, U.S. Public Health Service

Jane Preston, MD, President, Telemedical Interactive Consultative Services, Inc.

John Silva, MD, Program Manager, Advanced Research Projects Agency

Lawrence K. Vanston, PhD, President, Technology Futures, Inc.

Donald M. Vickery, MD, Chairman and CEO, Health Decisions International, LLC, University of Colorado, Health Sciences Center

Mae Waters, PhD, Comprehensive School Health Program, Office of Policy Research and Accountability, Florida Department of Education

John Wennberg, MD, Director, Center for Evaluative Clinical Sciences, Dartmouth Medical School

Barry G. Zallen, MD, Harvard Community Health Plan

Foreword

C. Everett Koop
Dartmouth University

Michael D. McDonald
The C. Everett Koop Foundation

Two revolutions—health system change and the building of the national information infrastructure—are transforming American health care and dramatically affecting the health of Americans. The specific shape and pace of this transformation are impossible to predict today, but our observations of early initiatives give us some modest confidence that present day concepts of what determines health and disease and our methods of intervening will be dramatically different in the not too distant future. It also appears that the emergence of an intelligent network will be a central organizing mechanism in the transformation of the health system.

In 1993, Vice President Gore asked the Koop Institute to help stimulate the private sector's leadership and involvement in the health component of the national information infrastructure, and to help foster public–private collaboration. Since then, we have carefully monitored the flood of initiatives on communication technologies for health care. Distinct and sometimes overlapping proposals have been put forward by all participants in the health care debate: the White House, Congress, health care professionals, health sector institutions, communication companies, computer firms, information providers, researchers, state and local government agencies, the media, the scientific community, and the general public. Such broad involvement shows the intense interest in properly harnessing the expanding technology of the communication revolution to transform the health care system.

Interactions between these revolutions will lead to cascading effects that are impossible to predict in their full complexity. Because health care is not the same thing as health, our focus is on early defining influences that help us forecast lasting effects on the health of Americans.

The tenets of chaos theory and the sciences of complexity remind us of the rule of first forces—that the first intervening events and influences are disproportionately influential in determining outcomes. Modeling first forces in the context of change helps the observer anticipate the emerging system.

As the nation has wrestled with new initiatives in health care and information infrastructure, we already see patterns of growth that form a useful basis for forecasting what an American health care system might look like by the year 2000. Yet the field remains a rich and uncharted frontier that beckons the scientist, the policy maker, and the entrepreneur to make critical contributions.

This book is the best compendium of these first forces, which will help determine the scope and potential of the emerging interactive media as they are being applied to health concerns. The distinguished authors, all pioneers in their own fields, describe such things as member-centered managed care, demand management, telemedicine, provider teamwork, patient involvement in health care decision making, reinventing government, new media pedagogy, interactive health education in schools, simulation in health education, and the new dynamics of public–private sector responsibilities.

For readers who are struggling to understand health from the perspective of the new media, or the new media from the perspective of health, this book will help them knit together the early vectors of managed care, a reinvented public health and health education, an empowered public, and the interactive media into a tapestry of their own making on which future contributions will be made.

Preface

This book sprang from a forum on interactive multimedia and health care held November 1992 in Washington, DC. The meeting was cohosted by the assistant secretary for health, head of the U.S. Public Health Service (PHS), and the deputy assistant secretary for commerce, head of the National Technology and Information Agency (NTIA). The conference was sponsored by the PHS, NTIA, the National Demonstration Laboratory for Interactive Information Technology at the Library of Congress, The Interactive Multimedia Lab at Dartmouth Medical School, The Center for Technology, Policy and Industrial Development at MIT, AT&T, Bell Atlantic, MACRO International, and PictureTel. Its purpose was to bring health policy makers and new media researchers together "to promote informed design, production, and use of multimedia for the promotion of health and prevention of disease." Rarely had representatives from these two domains interacted; the conference became a cross-cultural experience for many and, I believe, the breeding ground of new professional relationships.

There was, of course, a certain amount of enchantment in the conference air. Mitch Kapor, founder of Lotus Development Corporation, participated from his Cambridge, Massachusetts, office via interactive television. For some participants, this was their first experience of real-time interactive teleconferencing. For the old hands, it was a thrill just to see the technology, ever flirting with failure, come through "just in time." Multimedia presentations, stapled and scotch-taped together behind the scenes, actually dazzled as intended. And, on occasion, technology wizards and health policy wonks crossed language and cultural divides to proclaim awe for each others' work.

The Forum on Interactive Multimedia and Health Care was not really about the unbridled potential of new technologies, the kind of thing one might find at a large multimedia product show. Rather, the excitement came from sharing our hopes of harnessing the new interactive, connected, and user-driven media for improving the health of Americans. For even in 1992, before President Clinton's legislative push for health care reform, the goals were clear—to increase access, to improve quality, and to manage costs. Since the conference, these criteria have become the mantras of health care improvement; they continue to focus the enthusiasm of a growing number of new media developers.

A book on new media and health was suggested to me at the conference by one of its participants, a long time friend and colleague, Dr. Jennings Bryant. Jennings recommended a book that would foster and focus the cross fertilization between health specialists and new media developers so the new media could be put to work to improve health. The fostering function is served here by giving the authors, many of whom attended the conference, a larger forum in which to offer their work. The focus comes from the now familiar three health care improvement criteria by which these and all researchers and developers of new health media will ultimately be judged.

The book is divided into six sections: Overview, Delivery, Health Information, Health Education, Potholes Along the Information Superhighway, and a new media Glossary.

Overview

The overview juxtaposes characteristics of the new media (interactive, connected, and user driven) with three criteria for health care improvement: increased access, improved quality, and cost management. It offers a "new media and health matrix" of criteria for building and evaluating emerging health care systems.

Delivery

The section on how new media can enhance the delivery of health care includes chapters on: (a) managed care, (b) demand management and self-care, (c) telemedicine for rural residents, and (c) how the Internet can be used to facilitate collaboration among health researchers and providers.

Managed care is one of the most promising organizing principles in the health care reform movement. The Harvard Community Health Plan has initiated a pilot project using a telecommunication network that connects some of its members with the clinical information system used by the clinician at one its health centers. HMO members may interact with the system and providers from their home via minitel computers. This project has much to teach us about the role of the new networks, their media, and their efforts to involve providers, patients, and the well population in the health care process. Barry Zallen, provider and researcher at the Harvard Community Health Plan, discusses this project and his vision of the future of member-centered managed care networks.

Donald Vickery, the author of the best selling book, *Take Care of Yourself,* discusses how information technology helps manage unnecessary service demand, by helping patients take care of themselves. Informed self-care thus provides a counterpoint to the "services reduction" dimension of health care reform.

Interactive television is a promising medium through which isolated patients and providers can consult specialists. Jane Preston, one of the pioneers in telemedicine, describes how her Texas project successfully designed and tested a cost effective health care system in small, isolated rural communities. Preston's success is measured in terms of system reliability as well as user satisfaction.

In the Internet chapter, Anthony Gorry, John Silva, Joseph Eaglin, and I discuss ways in which advances in computing, networking technologies, and research methods can enhance the integration of health and human services. By linking health experts electronically, they can collaborate across time, space, and organizational boundaries. We advocate collaborative research that uses the Internet to form virtual laboratories (collaboratories) for the research, development, and implementation of health care technologies.

Health Information

Health information is the life blood of health care. Providers, patients, and the well population need timely and accurate information. In this section, two chapters address the potential for: (a) extending the traditional flow of health information (from researchers to providers) to reach patients who want to share in decisions about their care; and (b) the federal government's role in providing health information to the public.

John Wennberg, a noted expert in health outcomes research, has developed a set of interactive video programs to help doctors and patients share in decision making. These programs have been effective in reducing the use of unnecessary health care services. Wennberg and his team are working with kiosks as well as networked versions of their programs. He discusses his shared decision making model and the importance of an informed patient in improving the quality and cost effectiveness of health care. He shows how networking technology will enhance this effort.

Michael McGinnis, Mary Jo Deering, and Kevin Patrick, from the U.S. Office of Disease Prevention and Health Promotion of the U.S. Public Health Service, discuss the importance of prevention and health promotion in health care reform. They describe the current and future role of public health information services in supporting national health goals and the national information infrastructure.

Health Education

Each of the three chapters in the section on health education takes a different approach to the topic: (a) integrating multimedia health programming for public schools, (b) using networked multimedia and simulation technologies and new learning theories that promise to transform public health education, and (c) educating health providers and patients through interactive media and drama.

Sandra Cheiten, of a major broadcast network, in collaboration with Mae Waters, a leading health educator, describes how departments of education and local schools have worked with ABC News InterActive to develop and implement a series of interactive multimedia health education programs in the classroom. This series includes programs on AIDS, alcohol abuse, teenage sexuality, and drug and tobacco use. The authors emphasize the importance of comprehensive school health education and make recommendations for successfully integrating multimedia programs into the curriculum.

Network-based multimedia and distributed simulation are intriguing vehicles for health-related education; they are becoming practical through the evolution of America's national information infrastructure. Chris Dede and Lynn Fontana describe how their work with these emerging technologies, which center on new learning theories, will provide alternative approaches to health education and improving public wellness.

Joe Henderson has been active in professional and patient health education using interactive media and virtual environment technologies. He believes strongly in the role of stories, whether crafted or real, in helping people—provider or recipient—learn about the "swamp" of clinical practice. In this wide-ranging discussion, Henderson offers a provocative view of new media and education that is both optimistic and cautionary.

Potholes Along the Information Superhighway

A final chapter provides a sobering balance to otherwise rather optimistic assumptions that a national information infrastructure will be forthcoming.

Frank Fisher questions whether many of the potential health-promoting uses of new information media will soon be realized for all Americans. Effective health applications, he points out, require a switched broad-band national information infrastructure. He argues that free market competition among telecommunication firms, by itself, is not likely to make switching accessible to all Americans. Nor, despite federal health reform efforts, will market demand alone assure the development of health information products the public needs. He concludes that more government action, spurred in part by those concerned with the health of the nation, will be required to get health services rolling down the electronic information superhighway to all Americans.

The New Media: Annotated Glossary

Julia Marsh and Larry Vanston have compiled a glossary of computing and networking technology tools. For readers who are not fluent in cyberlanguage, the glossary alone is worth the price of this book.

I hope that readers will experience what the participants of the Forum on Interactive Multimedia and Health Care did—informed yet cautious encouragement to employ new media to increase access, improve quality, and manage the costs of health care.

ACKNOWLEDGMENTS

It is with great appreciation that I acknowledge the participants and planners of the Forum on Interactive Multimedia and Health Care and my fellow authors for their creativity, hard work, and continued indulgence in my obsession with juxtaposing seemingly unrelated ideas and people. I am especially grateful to Jennings Bryant, who always stays enough steps ahead of me to play a mentoring role and is generous enough to do so. To my recent colleagues at the Advanced Research Projects Agency, I am grateful for your patience with a health care policy wonk who secretly yearns to be a technology wiz. And to the members of the Health Information and Applications Working Group of the Information Infrastructure Task Force, thanks for being part of my continuing education program. Finally, deep appreciation goes to my own personal editor, Peter Ross Range.

Linda M. Harris

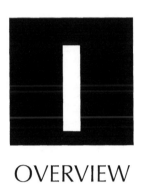

OVERVIEW

Differences That Make a Difference

Linda M. Harris
Center for Health Policy Research,
George Washington University

Though I do not believe that a plant will spring up where no seed has been, I have great faith in a seed. Convince me that you have a seed there, and I am prepared to expect wonders.

—Henry David Thoreau

You can count how many seeds are in the apple, but not how many apples are in the seed.

—Ken Kesey

Gregory Bateson, noted anthropologist and communication theorist, was fond of referring to "differences that make a difference." Bateson warned behavioral theorists not to confuse empirically based observations with systemic knowledge (Bateson, 1972). The same standard can help lay observers harvest pragmatic truths from the noise that sometimes poses as news.

There has been an extraordinary amount of media coverage lately about health care and the information age. Just as more is not necessarily better health care, more technology is not necessarily the answer to the health care dilemma. Yet telecommunication breakthroughs hold great potential if properly harnessed. One premise of this book is that the primary goal is not simply to improve the fiscal health of the existing medical systems, but to improve the actual health of individuals. Health care systems, like the emerging computing and networking tools supporting them, are the means to that goal.

This overview chapter offers a framework for evaluating health care systems and their new media tools. Just as there are measures of individual health, there are measures of health care systems' effectiveness in meeting their goals. Health care systems and the new media tools that maintain them, must adhere to objectives consistent with the primary goal.

The evaluation framework consists of the primary health goal, health care system objectives—access, quality and cost management, and characteristics of new media tools—interactive, seamlessly connected and user driven. Each chapter in this book provides a unique perspective on the goal, the benchmarks for success, and new tools to meet these objectives.

The book's final chapter serves as a counterweight by questioning whether the information infrastructure envisioned by the private and public sectors will provide the fertile ground needed for new health media systems to flourish. Together, the first and last chapters are intended to help us be critical observers as well as participants in the evolution of health care and its new media tools.

LEGACIES OF THE 1993–1994 HEALTH CARE
REFORM EFFORT

Almost one year from the day the Clinton administration first announced its health care reform plan in September, 1993, it was proclaimed dead in Washington, DC. But a year's worth of reform rhetoric and legislative infighting has left two important legacies—greater consensus on our goal and an expanded universe of stakeholders.

During the national debate, conservatives and liberals, corporate executives and public servants, providers and citizens, came significantly closer to a shared vision of the purpose of health care systems. That is, to improve the mental and physical health of everyone at an affordable cost (Field, Lohr, & Yordy, 1993). What was once a low income issue has become a growing concern for many Americans, regardless of economic class. Consensus on this goal is the first step toward its achievement.

Even after its efforts failed in the fall of 1994, the Clinton administration, along with states and private health care organizations, has forged ahead toward changes in the American health care system. And changes will doubtless be confirmed by future administrations. Quite apart from the success or failure of federal legislative efforts, progress is very much alive and gaining momentum in many states and private health care organizations. (Anders, 1994; Flower, 1994; Sands, 1994; Skolnick, 1994; Senate Committee on Finance, 1993; Tallon & Nathan, 1993). Morgan summarizes the dynamics of health care reform after the 1994 federal legislative attempt:

> Together, the pressures on state budgets and business bottom lines are changing medical care on a level hardly envisioned when Clinton unveiled his proposals. . . . States' embrace of "managed care" mirrors a transition that is

producing unprecedented changes throughout the private health care market place." (Morgan, 1994, p. A1).

Internal focus on the administrative complexity and inefficiencies in our current health care delivery system has engaged legions of computer engineers and networking specialists. The authors of this book represent hundreds of information and computer scientists and researchers who see themselves as stakeholders in the transformation of healthcare. Their work is now focused upon improving health information availability and useability.

Because they, for the most part, work outside the current health care system, the view of possibilities is not narrowed by the perspectives of traditional hospital administration or medical education. They are stakeholders in health. They see beyond the current boundaries of the medical system to the wide array of resources that can be brought to bear to improve health. Upon discovering that the vast majority of all frontline health decisions are made outside the doctor's office, these pioneers plan to bring health information to the customer. Some technologists see public health providers, with their emphasis on prevention and community services, as their partners in building an enhanced and customized health information infrastructure. Others are collaborating with primary and secondary schools to incorporate interactive health education into the curriculum.

Together, computer and network developers envision a more expansive, responsive, and cost-effective system of care than exists today.

However, technologists have not yet fully engaged the health policy community in an appreciation of how technology could advance the cause of health. One reason, perhaps, is that they tend to describe technology changes as either propelling us to some faster, fully computerized time (the information age), sending us to a strange and somewhat cosmic space (cyberspace), or supporting us with a new set of wired underpinnings (the national information infrastructure). Understanding these terms requires specialized inside knowledge that is inaccessible to most people, and this specialized knowledge offers little to help us become informed consumers or even advocates of their work.

One successful approach has been to observe changes in the information and communication tools with which we are already familiar so that the abstract becomes more concrete. With a critical eye, we see dramatic upheaval. The old media—the telephone, radio, television, the photograph—are being transformed in ways that are observable and measurable. And the new characteristics— interactive, seamlessly connected, and user driven— are the new standards by which we can evaluate their contribution to improved health for everyone at affordable costs.

In summary, the 1993-1994 national health care debate left a blueprint for action and a larger world of stakeholders. The authors of this book represent some of the stakeholders who will be building the health care tools of the future. To set the stage for their work we briefly review criteria by which to evaluate health care systems and the new media tools that support them. Finally, I offer a synopsis of each chapter in light of this evaluation framework.

Health Care System Objectives

Policy makers have reached consensus on three clear indicators of progress toward our goals of improved health for all at affordable costs (Institute of Medicine [IOM], 1993): access is increased, quality is improved, and costs are managed. An intervention in the health care system—including new media—must favorably impact at least one of these indicators to be considered effective.

Increased Access

The controversy over how to achieve universal health coverage obscured the almost unanimous agreement on the goal of universal access. However, coverage does not assure access. Almost all Americans have access to health coverage in some form or another either through a private or government insurance plan or by going to the emergency room in most hospitals. With new computing and networking technologies on the near horizon, we can think about access in terms of moving health expertise rather than moving people. This means that both health information as well as health services can be brought to the customer, removing many of the current barriers to access.

Ironically, health care reform efforts may bring us "back to a future" where individuals and families reclaim a central role in caregiving (Harris and Hamburg, 1990; Starr, 1982). Patients and the well population make 80% of their health-related decisions outside a medical setting (Sobel, 1987). Their ability to navigate the health care system, select appropriate care, and prevent unnecessary health problems, depends upon how well-informed they are. A group of leading health care specialists agreed that access to health information should be a part of every American's right to health care (Carlson et al., 1992). A 1993 Institute of Medicine (IOM) report addressed the need for "broad public health and health education initiatives that help people understand how to take care of their health, use health care services appropriately, and seek healthful environments in the home, the workplace and the community" (p. 15).

The pragmatics of access often have a lot to do with how close and how user-friendly a health care service is: How inconvenient does a health care service have to be to count as inaccessible? Two bus transfers and a taxi ride to the nearest prenatal care clinic, followed by a 2-hour wait can be a powerful disincentive for a poor, pregnant woman. A 5-hour drive from a rural community to the nearest big city hospital can dissuade a farmer from seeking care even as his health deteriorates and the cost of delayed care mounts.

Dena Puskin, Deputy Director of the Office of Rural Health Policy, sees an important role for telecommunications in making care more accessible to isolated populations

> Too few primary care practitioners and the need to travel long distances for specialty care have made it difficult for many rural residents to receive the care they need when they need it. It is now technologically possible for rural patients to have consultations with distant specialists without leaving their communities. (1994, p. 1)

Improved Quality

Although most readily agree that more health care is not always better health care, we have relatively recently become acutely aware of the variations in the quality of care (Wennberg & Gittelsohn, 1982). An IOM reform committee identified two major changes in medical care that are expected to dramatically reduce unwarranted variations in quality of care: (a) "Care is being evaluated increasingly on the basis of its process and outcomes; and (b) with the advent of better research methods and computer technology, clinical medicine is becoming more science- and information-based." (Field, Lohr, & Yordy, 1993, p. 33).

By providing analyzed clinical outcomes data and clinical practice guidelines at the point of service delivery, patient and provider decisions can be based on the best information available. With this information in hand, providers and patients can make more informed and timely decisions concerning the efficacy and risks of specific health interventions, including necessary treatment and possible prevention. However, too much information can delay proper treatment. Information retrieval tools are needed to manage the overload of health information that swamps providers and confuses patients.

The relatively recent discipline of outcomes research is generating a systematic process by which clinical practice guidelines can be developed and disseminated to assist practitioner and patient decisions about appropriate health care. These guidelines cover treatment as well as preventive services such as early diagnosis, screening, immunizations, and counseling (Field & Lohr, 1990,1992; U.S. Preventive Services Task Force, 1989). Guidelines can contribute to the improvement of quality assurance while reducing the medical liability for bad clinical outcomes (Gelijns & Halm, 1992). The effectiveness of such a guidelines process depends upon a robust and ubiquitous information infrastructure.

The high quality of medical education in the United States has long been associated with the high quality of care we enjoy. Over time, medical education has emphasized specialized medicine over primary care. Yet, the cost of educating in and paying for specialty care can no longer be sustained. A number of recent IOM studies have called for increased emphasis upon primary care, nursing, and allied health education (1978, 1983, 1989). Telecommunication can play an important role in distributing professional education throughout the country, especially to traditionally underserved areas. It can also be used to support real-time training of primary care and allied health professionals by consultation with specialists via interactive television. Primary care will be more viable when specialist help can be invoked by telecommunications.

Quality health care has always been associated with a high level of confidentiality between provider and patient. Any information technology used to support provider-patient interaction will have to sustain the same level of privacy. There is a great deal of activity surrounding this issue in the private and public sectors. The Institute of Medicine, among others, has made confidentiality and privacy a high priority.

> Plans that rely heavily on health-related, patient-identified information in large databases must acknowledge the social, legal, and ethical problems, inherent ethical issues, and ideally propose some steps to protect patients' privacy rights. These safeguards must be strong (and must be perceived to be strong), but they should not interfere with appropriately approved research and system evaluations. (Field & Lohr, 1992, p. 61)

President Clinton established an interagency Information Infrastructure Task Force to address confidentiality, privacy, and security issues. Some of the solutions under review include developing mechanisms for security such as digital signatures and encryption capabilities, and establishing uniform privacy protection principles.

The participation of informed patients in defining quality of care and in the selection of treatment and preventive services may not only lead to greater patient satisfaction with the quality of their care but also reduce the use of unnecessary services (Leaf, 1993; McGinnis, 1992; Somers, 1984). As health information and decision support tools become computerized, patient participation in the health care process will be increasingly determined by how well they are connected to the information highway.

Managed Costs

The reduction of administrative costs through improved information systems has been well documented. (Kunitz and Associates, 1994; Tierney, Miller, Overhage, & McDonald, 1993; Workgroup for Electronic Data Interchange Report, 1992). Time and money are wasted navigating the "nonsystem" of health care that exists today. Overspecialization, with all of its benefits to medical science, has left providers and consumers in a maze that is almost impossible to wade through. The 1993 IOM reform report proposes that "well-managed, 'seamless' systems that integrate an array of medical services can improve the quality and efficiency of care" (p. 25).

The 1993 IOM report (Field, et al.) focuses most of its cost management proposals on resource use and health outcomes. Information regarding price, quality and expected outcomes "helps practitioners, patients, and others learn how actual care conforms to criteria for appropriate care and why care varies in effectiveness and efficiency" (p. 23). Decision support concerning appropriate services has also reduced costs in some preliminary studies (Evans, Pestotnik, Classen, & Burke, 1993).

The Health Project Consortium, a public and private sector team of health policy experts, business leaders, health insurers and government officials, agreed that broad access to health information services helps meet cost management goals by reducing demand (Fries, Koop, Beadle, et al., 1993). They maintain that consumer access to health information services would "reduce the burden of illness and thus the need and demand for medical services" (p. 3221). A growing body of evidence, gleaned especially from employer programs and managed care organizations, supports their case (Golaszewski, Snow, Lynch, Yen & Solomita, 1992; Lorig, Kraines, Brown & Richardson, 1985; Morrison and Luft, 1990; Vickery, et al. 1993).

The IOM report (Field et al., 1993) also supported the premise that information systems are essential to meeting the goal of improved health for everyone at affordable costs:

> Successful implementation of health care reform will require more and better data and information about health care, especially in the face of the pressures that can be expected as a new system tries to hold down expenditures while expanding access and maintaining high-quality care. Health care providers, patients, the public, and policy makers will be asked to make harder and more complex choices and trade-offs than in the past. Informed choices dictate a vastly increased need for improved data and information for operations, evaluation, and research. (pp. 58–59)

Two recommendations from the IOM report reflect the importance of information systems: (a) universal implementation of computerized patient records (CPRs) and CPR systems; (b) an expanded program in information services for health services research and technology assessment. (pp. 60–61)

In essence, these three health care system objectives can be linked: increasing access and assessing quality care are integral to cost management; this, in turn, can lead to increased access. Harnessing new media to this synergy may enhance the capacity of health care organizations to achieve the primary goals.

NEW MEDIA

In his influential book, *Technologies of Freedom,* Ithiel deSola Pool (1983) characterized the communication evolution. The first communication era, he wrote, was speech; the second, writing; the third, printing; and the fourth an era in which all media are becoming electronic.

Observing old, familiar media—television, radio, phone, print, photographs—change into new electronic media engenders recognition of the profound differences in how we manipulate information and communicate. The media with which we are familiar are becoming computerized and connected over networks; they are metamorphosing into something new—again. Perhaps we are in transition between the 4th era and a 5th era in which media are changing from passive, closed, and producer driven to interactive, connected, and user driven. Such profound changes, along with an historic shift in the locus of control to all users of the new technologies could make significant differences in how health care is provided and used.

Interactive

Computerized and digitized media can be made smart.

Once computerized, passive media can turn into interactive media—media that can not only inform their users, but interact with them. They can retrieve and compute an infinite number of variables and package the results in the format the user finds most meaningful. Computerized media, through their interactive capacities, can customize information according to each user's learning style, pace, and preferences. Interactive media can get to know their user's unique information needs, even anticipate them. They can serve as information brokers, matching information resources with information users throughout a network of users.

Interactive televisions, for example, can monitor the user's information demands. They can learn to anticipate what we want to know and in which medium we most want to receive new information, then search the "world wide web" of networked information sources. They can provide immediate feedback in multiple media—printed data, video, graphics, or in animated simulation. Using feedback about a user's health care options, these media can provide decision support by simulating future scenarios in which providers or patients can play out their options without suffering the consequences of choosing badly.

Seamlessly Connected

Old media typically originate from a single source (a publisher or a network) and distribute the same, one-way messages to mass audiences. These media have been relatively unconnected; televisions have not transmitted phone calls, computers have not collected the *CBS Evening News*. Instead, they have comprised a series of separate information "stovepipes" reaching into our homes, businesses, schools, and hospitals.

By digitizing information into ones and zeros all media become mutually translatable and transferable. "The content becomes totally plastic—any message, sound, or image may be edited from anything into anything else" (Brand, 1988, p. 18). The new digitized media, carried over switched networks or wireless transmissions, can carry messages back and forth from anyone who is part of a web of connections. These media cast messages over the network broadly (to every user) or narrowly (to a select few). Adding intelligence to an interactive set of media increases the information processing capability of everyone on the network. Decentralized, intelligent media will mean that people at home, at school, or at work can combine the capabilities of broadcasting and computing for their own information and communication purposes. Self-help groups, home care, and personal health information services can thrive.

The opportunities for seamless connectivity abound—as do the barriers. Samuelson (1994) describes the opportunities,

> The advent of fiber-optic cables, digital switching and new wireless (radio) transmissions is breaking down barriers between [sic] local phone companies, cable companies and long-distance companies. Firms are scrambling to merge or create partnerships what would ultimately allow them to offer customers a full array of services. (p. 27)

There are still media vendors building proprietary tools that do not communicate with other tools. However, the government is becoming more aware of ways to provide incentives to those who will build from an open architecture (see Principles of the NII, in chapter 7). The need for, as well as the barriers to, seamless, universal connectivity are addressed by Fisher in the final chapter of this book.

User Driven

Once computerized and carried over an open set of networks, information can be produced almost anywhere, conveyed in any medium, over any network and stored almost anywhere. The old distinctions between publisher and reader are blurred; everyone on the network is a user. Users of open networked information can integrate multiple sources, forms, and media to construct the most meaningful message.

The new, user-driven media have given rise to a new human–computer interaction research paradigm. Researchers and developers across multiple disciplines are devoted to improving the manner in which people can interact with the computer by enhancing user-friendliness and efficiency. For example, the Advanced Research Projects Agency, of the U.S. Department of Defense, has a research program dedicated to developing various interfaces, including speaker independent voice recognition tools. A key objective is to make the computing tools transparent to the user. This research effort will help place smart multimedia at the command of people who know or care little about the inner workings of those media.

In summary, new health media tools promise customized, coordinated, and participatory care. These promises can be judged according to whether they help achieve universal access to health services and information. They can be evaluated according to how they improve the quality of the service delivery and how comprehensive, timely, and well distributed the decision support and educational services are. Finally, new health media tools can be judged by whether they contribute to our ability to manage costs.

A VIRTUAL VISIT

The written word cannot fully capture the profound difference new media developers are trying to make in health care. To experience the effort personally, readers are invited to visit a virtual health care telecommunication system funded by the Advanced Research Projects Agency and the National Science Foundation. It is called the Networked Multimedia Information Service (NMIS). On-line visitors enter through the Internet's gopher far.mit.edu or the World Wide Web Mosaic service at: http://www.nmis.org. There they will find the National Multimedia Information Service Home Page.

This pilot project in its early stages, is cosponsored by the Massachusetts Institute of Technology, Dartmouth Medical School, Carnegie-Mellon University, IBM, and Turner Broadcasting Company. A research team is developing and implementing a prototype of an interactive health media system that opens traditionally closed media by connecting them over the internet (Massachusetts Institute of Technology, 1993).

This experimental service includes the *CNN Newsroom*, the award-winning, noncommercial cable news service to the classroom developed by Turner Educational Services. By placing CNN on an interactive network, its viewers become users. On-line CNN users will be able to browse CNN archives and select news clips of greatest interest to them.

By connecting CNN to Dartmouth Medical School's multimedia server, users will be able to reach into the medical school's health information archives and retrieve the information most useful to them. Through access to MIT's server, users will also be able to access the software programs that allow them to combine news clips with health information for their own information and communication purposes. In the case of currently

FIG. 1.1.   Networked multimedia information services home page.

underserved providers and citizens, these gateways to various media allow them to keep up with the latest medical news and practice guidelines, or participate in training otherwise unavailable to them.

During a virtual visit to the NMIS users will find developers who share a long-range vision of interactive, seamlessly connected and user-driven media, contributing to integrated, customized, participatory care on demand with ubiquitous reach. But these researchers are far from realizing their vision today. Rather, they are on the leading edge of new health media research—an exhilarating place where there are still more questions than answers.

NEW MEDIA AND HEALTH

It is still too early to predict the critical path to achieving systems that offer high quality and affordable care to everyone, but the work in this volume projects us into the near future of health care organizations.

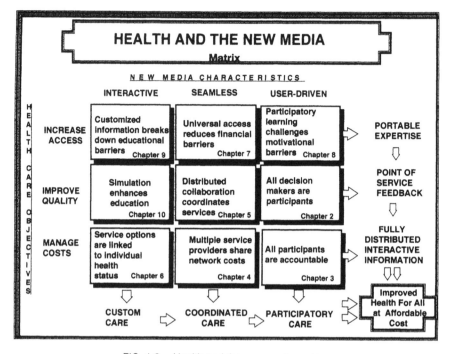

FIG. 1.2.   Health and the new media matrix.

At the risk of characterizing their work too narrowly, I have placed each author's chapter in one of the cells of the health and new media matrix, juxtaposing new media characteristics with health care system objectives. (Several sets of authors could be in a number of the cells). An examination of each cell gives us a preview of the emerging health media tools.

Access and the New Media

Three chapters highlight access by making expertise portable. Education is a significant barrier to appropriate care and information about self-care. Many people with low literacy skills suffer disproportionately from preventable or manageable conditions. Dede and Fontana (ch. 9, this volume) illustrate, among other things, how interactive, simulated health education programs can make health expertise more accessible to low literacy populations.

The federal government currently assumes responsibility for assuring universal access to existing telephone networks. In chapter 7, McGinnis, Deering, and Patrick offer a rationale for extending the role of the federal government to assure new health media especially where financial barriers exist.

Individual responsibility for maintaining one's health must begin early, before unhealthy habits and lack of motivation become barriers to making responsible use of health care services and healthy lifestyle choices. Cheiten and Waters (chapter 8) describe the integration of interactive health education into early classroom experiences.

Quality and the New Media

New media can improve health care quality by linking research, practice, and education in the same feedback loop that extends to the desktop of the provider and the patient. Smart media can extend the education of health professionals. The multimedia programs Henderson describes (chapter 10) can be used in medical schools to simulate real practice experiences. They can also bring "just-in-time" training and expertise to traditionally underserved areas.

Interdisciplinary research and practice, conducted over a distributed network, can improve the quality of care in at least two ways. Gorry, et al. (chapter 8) describe how providers can be involved in the development of the tools they use, making them more functional and relevant to the task. They also describe ways that services can bring better coordinated and integrated care to patients.

Harvard Community Health Plan (chapter 2) participants have a media system that enables providers and patients to participate in the care process from their homes. Zallen discusses how quality of care can be assessed and improved upon at the point-of-service delivery by all participants in the system.

Cost Management and the New Media

Cost management is in large part a result of increased access to quality care and thus implied in all the chapters. Fully distributed interactive information systems hold promise for achieving this objective. Three authors have documented the cost management implications of their work. Wennberg (chapter 6) has built an interactive decision support system that personalizes a set of service options according to the condition of a particular patient. Patients who share in making service delivery options with this support have proven to be more prudent users of the health care system.

Preston (chapter 4) has built and tested a networked teleconsultation system. She demonstrates the financial savings of such a system and argues that the real cost savings will occur when communities share the cost of *connecting* multiple services. In order for these benefits to be realized, an open architecture is required.

In chapter 3, Vickery explains why and how individuals, by assuming more responsibility for becoming informed about their own health and health care needs, can reduce the demand for unnecessary services. As the human-computer interface becomes friendlier and more accessible from home, these costs are expected to drop even more dramatically while individual health status rises.

Conclusion

Health care reform is filled with both the promise and hyperbole of information age rhetoric. The authors of this book and I have tried to separate the wheat from the chaff. The writers offer tools that promise to make a difference in our personal and collective well-being; I suggest a set of conceptual tools for holding all of us to that promise.

REFERENCES

Anders, G. (1994, July 8). States slow ambitious health-care reform moves but continue to act as laboratories of innovation. *The Wall Street Journal,* p. A12.

Bateson, G. (1972). *Steps to an ecology of mind.* New York: Ballantine.

Brand, S. (1988) *The media lab: Inventing the future at MIT.* New York: Penguin.

Carlson, R. J., Ellwood, P. M., Etzioni, A., Goldbeck, W. B., Gradison, B., & Johnson, K. E. (1992). *Healthy people in a healthy world: The Belmont vision.* Alexandria, VA: Institute for Alternative Futures.

Evans, R. S., Pestotnik, S. L., Classen, D. C., & Burke, J. P. (1993). Development of an automated antibiotic consultant M.D. *Computing, 10,* 17–22.

Field, M. J., & Lohr, K. N. (Eds.). (1990). *Clinical practice guidelines: Directions for a new program.* Washington, DC: National Academy Press.

Field, M. J., & Lohr, K. N. (Eds.). (1992). *Guidelines for clinical practice: From development to use.* Washington, DC: National Academy Press.

Field, M. J., Lohr, K. N., & Yordy, K. D., (Eds.). (1993). *Assessing health care reform: Committee on assessing health care reform proposals.* Washington, DC: National Academy Press.

Flower, J. (1994, January). The other revolution. *Wired, 2.01,* 108–150.

Fries, J. F., Koop, C. E., Beadle, C. E., Cooper, P. P., England, M. J., Greaves, R. F., Sokolov, J. J., Wright, D., & the Health Project Consortium (1993). Reducing health care costs by reducing the need and demand for medical services. *New England Journal of Medicine, 329*(5), 321–325.

Gelijns, A. C., & Halm, E. A. (Eds.). (1992). *Medical innovation at the crossroads. Vol. 2. The changing economics of medical technology.* Washington, DC: National Academy Press.

Golaszewski, T., Snow, D., Lynch, W., Yen, L., Solomita, D. A. (1992). Benefit-to-cost analysis of a work-site health promotion program. *Journal of Occupational Medicine, 34,* 1164–1172.

Harris, L. M., & Hamburg, M. A. (1990). Back to the future: Television and family health-care management. In J. Bryant (Ed.), *Television and the American family* (pp. 329–348). Hillsdale, NJ: Lawrence Erlbaum Associates.

Institute of Medicine. (1978). *A manpower policy for primary health care.* Washington, DC: National Academy of Sciences.

Institute of Medicine. (1983). *Nursing and nursing education: Public policies and private actions.* Washington, DC: National Academy of Sciences.

Institute of Medicine. (1989). *Allied health services: Avoiding crises.* Washington, DC: National Academy of Sciences.

Institute of Medicine (1993). *Assessing health care reform.* Washington, DC: National Academy of Science.

Leaf, A. (1993). Preventive medicine for our ailing health care system. *Journal of the American Medical Association, 269,* 66–68.

Lorig, K., Kraines, R. G., Brown, B. W., Jr., & Richardson, N. A. (1985). Workplace health education program that reduces outpatient visits. *Medical Care, 23,* 1044–1054.

Kunitz and Associates (1994). *Final Report* (Contract 282-91-0062). Washington, DC: Agency for Health Care Policy and Research.

Massachusetts Institute of Technology, Dartmouth College and Medical School, Carnegie Mellon University, Turner Broadcasting-Turner Educational Services, Inc., International Business Machines Corporation. (1993). Research in networked multimedia information services. *Proposal to the National Science Foundation and Advanced Research Projects Agency.* Cambridge: MIT Press.

McGinnis, J. M. (1992). Investing in health: The role of disease prevention. In R. H. Blank, & A. Bonnicksen (Eds.), *Emerging issues in biomedical policy: An annual review, Vol. 1* (pp. 13–26). New York: Columbia University Press.

Morgan, D. (1994, Sept. 28). Fundamental change rippling through state governments. *The Washington Post,* p. A1.

Morrison, E. M., & Luft, H. S. (1990). Health maintenance organization environments in the 1980s and beyond. *Health Care Finance Review, 12* (1), 81–90.

Pool, I. D. S. (1983). *Technologies of freedom: On free speech in an electronic age,* Cambridge: Harvard University Press

Puskin, D. S. (1994). *Reaching rural: rural health travels the telecommunications highway.* Rockville, MD: Federal Office of Rural Health Policy.

Samuelson, R. J. (1994, Sept. 23). The merger urge. *The Washington Post,* p. A 27.

Sands, A. (1994). *Survey of health care reform: Fifty states & Washington DC.* Upland, PA: Diane.

Sckolnick, L. B. (Ed.). (1994). *Health care reform: State profiles.* Leverett, MA: Rector.

Senate Committee on Finance. (1993, June 15). *State health care reform.* (Item No. 1038-A). Washington, DC: U.S. Government Printing Office.

Sobel, D. S. (May 1987). In A. H. Levy & B. Williams (Eds.), Self-care in health: Information to empower people. *Proceedings of the American Association for Medical Systems and Informatics, Congress 87, San Francisco, CA.* (pp. 12–125). Washington, DC: American Association for Medical Systems and Information.

Somers, A. R. (1984). Why not try preventing illness as a way of controlling medicare costs? *New England Journal of Medicine, 311,* 853–856.

Starr, P. (1982). *The social transformation of American medicine.* New York: Basic Books.

Tallon, J. R., Jr., & Nathan, R. P. (1993, Winter) A federal/state partnership for health system reform. *Health Affairs,* 7–16.

Tierney, W. T., Miller, M. E., Overhage, J. M., & McDonald, C. M. (1993). Physician inpatient order writing on microcomputer workstations: Effects on resource utilization. *Journal of the American Medical Association, 269,* 379–383.

U.S. Preventive Services Task Force (1989). *Guide to clinical preventive services: An assessment of the effectiveness of 169 interventions.* Report of the U.S. Preventive Services Task Force. Baltimore: Williams & Wilkins.

Vickery, D. M., Kalmer, H., Lowry, D., Constantine, M., Wright, E., & Loren, W. (1983). Effect of a self-care education program on medical visits. *Journal of the American Medical Association, 98,* (250):2952–2956.

Wennberg, J., & Gittelsohn, A. (1982). Variations in medical care among small areas. *Scientific American, 246,* 120–134.

Workgroup for Electronic Data Interchange. (1992). *Report to Secretary of U.S. Department of Health and Human Services.* Washington, DC: U.S. Department of Health and Human Services.

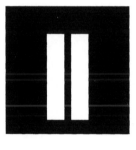

DELIVERY

Member-Centered Managed Care and the New Media

Barry G. Zallen
Harvard Community Health Plan

THE WORLD'S BEST PHONE MESSAGE

Daniel F. and his mother are members of the Harvard Community Health Plan (HCHP) where I practice medicine. Recently, Daniel's mother used her home terminal to get some advice about Daniel's cough. She answered several questions on her screen and was told that I would call her to further discuss his condition. Simultaneously, we were alerted at the office and I found "the world's best phone message" about Daniel on my terminal. All of his mother's responses and other significant information from his medical record were organized for me on screen. In my conversation with Daniel's mother she didn't have to repeat anything she'd already elaborated on. I found all the information from his record that I needed right on screen (without having to look it up), such as whether he has asthma or a drug allergy. Fortunately, he only had a cold and I gave some advice about treatment.

On another occasion, Daniel's mother used her terminal to get advice when he was vomiting. She answered all the questions, and was advised that Daniel appeared to have viral gastroenteritis ("the stomach flu"). Further, she was offered advice from the health education library about how to help him and what to expect. Daniel recovered easily and his mother had all the support, information, advice and reassurance she needed without ever speaking to a clinician

Both of these episodes were automatically recorded in Daniel's medical record.

INTRODUCTION

Health care has been undergoing significant reform for the past 10 years. The major shift that has occurred is from a fee-for-service payment system to a managed care system. In fee-for-service medicine, doctors and hospitals are paid for each service they perform, such as an office visit or a day in the hospital. In managed care, a primary care doctor is responsible for all the health care needs of a patient. The doctor may be employed by a health care organization or may receive a certain amount of money for each patient (*capitation*). In both cases, they manage the patient's care by emphasizing preventive measures or ordering necessary tests and procedures, consulting with certain cooperating specialists, and avoiding inappropriate testing, treatment, and hospitalization.

The managed care model has become attractive to those who are trying to achieve quality care without ever increasing cost. The emphasis of managed care on keeping patients well, coordinating their care, and providing services that have proven their effectiveness, invite various uses of the new media. From patient education and information to outcomes research, the new media promise to further enhance the effectiveness of managed care systems for providers and members. This chapter discusses a managed care plan that has been experimenting with the use of computer and networking technology (Martin, 1993; Zallen, 1993). The goal of our effort is to produce a comprehensive clinical information system incorporating an automated medical record, automated test and consultation ordering and tracking, automated prescribing, and systems to provide health education information and medical advice to members in their homes.

MEMBER-CENTERED CARE

Harvard Community Health Plan is a 25-year old-health maintenance organization (HMO) which was founded as a staff model HMO and has been committed since its inception to innovations in delivering health care and to community service.[1] The plan currently has almost 600,000 members; the majority using freestanding, multispecialty health centers. These incorporate practices in internal medicine, pediatrics, medical and surgical subspecialties, mental health, and obstetrics and gynecology. The centers also include laboratory, radiology, and pharmacy services. My center began serving patients in 1991, and we have been a test site for a new clinical information system since we opened.[2]

The HCHP places our members at the center of the health care process,

---

[1]Just a few examples of the commitment to community service include the work of the HCHP Foundation which has supported activities such as an interactive exhibit about lifestyle choices and health consequences at Boston's Museum of Science ("Ben's Grille"); the development of the "Talk, Listen, Care Kit" to aid families in discussions about drugs and sex; and the Foundation has helped to support the volunteer efforts of HCHP staff working on the "Alliance for the Homeless."

[2]The Robert H. Ebert Burlington Health Center of HCHP opened in January 1991. The

providing services to our members before, during, and after health problems occur. The new media technologies we are using at the center include basic computer-assisted technologies and an interactive network. By automating information we can provide more timely and comprehensive services, and better coordinate a patient's care than when we depended upon a paper-based information system. Connecting our providers and members to an interactive network creates a more intimate partnership between patient and physician, meeting the patient's needs more effectively and efficiently and with more focused use of a physician's skills.

Health Information to the Home

It is estimated that about a third of clinical visits are for information that could have been provided in a more meaningful, convenient, and cost-effective manner. For instance, patients often desire explanations about a minor illness or reassurance that an illness is not serious. In addition, they often want some advice about how to manage a relatively minor illness. Additionally, individuals and families are increasingly interested in information about how to maintain their health through diet exercise patterns, and changes in lifestyle habits. Successfully addressing these various needs without a visit to the doctor would contribute to *demand reduction* (Vickery, this volume, ch. 3). Studies have shown that even the use of telephones can meet patient needs and avoid travel for them and unnecessary visits at a busy practice (Wasson et al., 1992). Information to the home also helps members assume a more equal role in sharing the decision making process (Wennberg, this volume, ch. 6).

The Triage and Education System

Our effort to provide health information to our members in their homes, called the *Triage and Education System*, was intended to address these opportunities. Most families used a video display terminal connected to a standard telephone jack at the members home. It has a small screen, built-in keyboard, and a built-in modem. It is similar to devices that are present in almost every household in France where they are used for activities such as shopping and banking. Some families accessed the system through their personal computer and a modem.

Center, as of May 1994, serves approximately 14,000 patients. The Center has been the test site for a new clinical information system, InterPractice Systems(IPS), since it opened.

InterPractice Systems was founded as a joint venture of HCHP and Electronic Data Systems (EDS). The system was developed by staff from InterPractice Systems in collaboration with many clinicians and staff at HCHP, who contributed their expertise over several years as the system was developed and then tested at the Burlington Center.

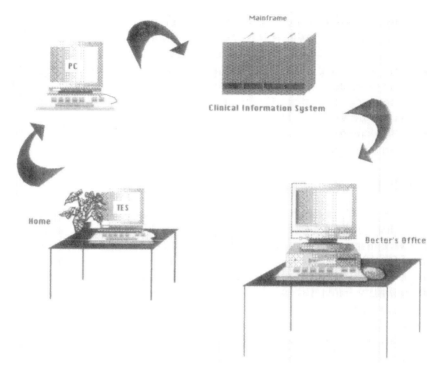

FIG. 2.1.   The interconnections between the patient's and doctor's terminals using the triage and education system (TES).

The home terminal, or personal computer through the modem and phone line, connects to several personal computers at our health center which, in turn, are connected to our main clinical information system computer. The personal computers gather information from the patient's medical record in the main computer when helping a patient and the interaction with the patient is documented in their medical record (Fig. 2.1).

Our members can access the following services from their home terminals: (a) health education library, (b) a triage system (illness protocols), (c) initial health assessment and health risk appraisal, (d) health highlights, and (e) suggestion box.

The health education library includes more than 250 topics with wide breadth and variety. For instance, there are modules on fat and cholesterol, exercise, smoking cessation, coughs, headaches, fever, contraception, vomiting, and vaginitis (For a full list of topics, see Appendix A.)

Members are able to browse at their desired level of detail and the information is accessible as modules. For instance, a family may wish to look up information about diet, but they may be most interested in fat and cholesterol. Therefore, they would select that module. Within a module they are offered different levels of detail and submodules may be linked to submodules in other topics. For instance, information about colds may offer linkages to "how to take a temperature," or "how to manage a fever."

A particularly important feature of the health education library, as with all aspects of our care, is that members are able to have unlimited access to the library at any time of the day or night without having to ask anyone or leave a message for anyone.

The triage system is intended to provide answers to questions about mild illnesses. The system is intentionally very conservative and was the product of extensive and multiple reviews by clinicians within their particular specialties. Members may choose to use the system in the *play mode*; no information from the interaction is recorded and there is no contact with our office. In the *real* mode, the interaction is documented and contact with us may occur. (For a complete list of protocols, see Table 2.1.)

When patients select a particular topic, they are asked a series of questions. The questions first relate to the most serious issues. For instance, if a patient chooses cough in a child (e.g., "Tommy") as the illness they are inquiring about, then the first questions would be "Is Tommy having an extremely difficult time breathing? Does he have any of these problems . . ." (Fig. 2.2). Whenever a patient (or parent) gives an answer that is outside the conservative guidelines designed by the clinicians, then they are instructed that they need contact with our office. The contact may take different forms.

TABLE 2.1
Illness Protocols

Adult Protocols

Back pain
Genito-urinary
Headache
Health risk appraisal
Initial health assessment
Muscles, bones and joints
Skin
Sleep problems
Upper respiratory infections
Vomiting, diarrhea, and abdominal pain

Pediatric Protocols

Fever
Initial health assessment
Upper respiratory infections
Vomiting, diarrhea, and abdominal pain

```
====================> TES <====================
First things first ...                        |
                                              |
Is Tommy having an extremely difficult time    |
breathing? Does he have ANY of these          |
problems:                                     |
                                              |
     *Gasping for air or having trouble        |
      getting enough air in and out           |
      (NOT due to stuffed up nose)            |
     *Being extremely agitated and needing     |
      to sit up to breathe                    |
     *Poor color                              |
                                              |
     [ ] A. Yes                               |<F1>Help
     [ ] B. No                                |<F2>Jump Back
                                              |<F3>
                                              |<F4>
                                              |<F5>
                                              |<F6>
                                              |<F7>
Press <X> to select answer, <Space> to erase, |<F8>Main Menu
<Enter> when done.                            |
                                              |
```

FIG. 2.2.   First question in cough protocol in triage and education system (TES).

If a patient's answer to a question falls into a severe category, then they will be told to hang up and that we will contact them immediately (Fig. 2.3). The support staff at our office is alerted to this emergency by an alarm. Our staff would then immediately alert a clinician and the information that was already formatted by the system would be available for the clinician to quickly review. We then are able to contact the patient and make the appropriate arrangements (for instance, sending an ambulance) given their condition.

Fortunately, such an occurrence is quite rare. The warning that patients are given when they first log on to the system indicates that they should not be using the system if they are concerned about a serious illness or have a serious complaint. In those cases, of course, they should be calling us directly, day or night. If a patient feels that their condition is not serious, then they are invited to proceed with the program.

Nevertheless, patients may not realize how ill they are and that some of the symptoms or signs they have would be of greater concern to us. By going through the program, their answers will indicate whether they need urgent contact. If so, it is arranged promptly.

More often, patients will proceed with the questions and answers and may give an answer that indicates a somewhat less severe situation. They will also be told that they need contact with us and that we will call them back within an hour. Again, our staff is alerted at the office (though not with an alarm) and the information is available for the clinician to review and then contact the patient.

Less urgent contacts may also be recommended as patients give answers that are less and less troublesome. Thus, some patients will be told that we will be getting back to them within a few hours and others are told that we need to see them that day. In all these cases, we are alerted at the office and will contact the patient or make arrangements for an appointment.

The answers patients give to the questions asked by the system and the information extracted from their records (such as current medications the patient is using) is formatted into positive findings (such as fever and sore throat) and negative findings (such as the absence of headache). This data can then be reviewed by the clinician. This synopsis of the patient's significant past history, medication history and significant findings during the current illness is what I like to call, "The world's best phone message". With this in hand (so to speak) we can call a patient and address their needs in a

```
==================>TES<====================

Michael, because you have a serious        |
problem getting your breath, we think      |
you should talk to a clinician immediately |
                                           |
                                           |
We will have someone call you right away   |
                                           |
Be sure to call the center at 617-221-2600 |
if you don't hear from us in 5 minutes.    |
                                           |
                                           |
                                           |<F1>Help
                                           |<F2>Jump Back
                                           |<F3>
                                           |<F4>
                                           |<F5>
                                           |<F6>
                                           |<F7>
Press <Enter> to continue                  |<F8>Main Menu
                                           |
```

FIG. 2.3.   An immediate call outcome for cough protocol in triage and education system (TES).

```
==================>TES<==================

Sherry, because Christopher has had a          |
temperature of 100.4 or more for 1 day         |
or less and no signs of a serious illness,     |
we think you can take care of his              |
problems by yourself at home.                  |
                                               |
Please be patient...advice for this is         |
coming up in a moment                          |
                                               |
                                               |
                                               |
                                               |
                                               |<F1>Help
                                               |<F2>Jump Back
                                               |<F3>
                                               |<F4>
                                               |<F5>
                                               |<F6>
                                               |<F7>
Press <Enter> to continue                      |<F8>Main Menu
                                               |
```

FIG. 2.4.   A self-care outcome for fever in a child in triage and education system (TES).

very focused and prompt fashion without having to ask them to review any of the information again. This is very satisfying for our members and for our clinicians.

In most cases, the answers that a patient gives during a protocol do not trigger any need for patient–clincian contact. For instance, a patient with cold symptoms may go through the protocol indicating a low enough fever, a mild enough cough, and a lack of other serious symptoms. A contact will not be generated and at the end of the program the patient will be informed that their symptoms appear to indicate that they have an upper respiratory tract infection (a cold). They are then offered information from The Health Education Library about what colds are and how to manage the symptoms. An example from the fever protocol can be seen in Fig. 2.4. If they wish to pursue the offer of information, they are then connected automatically with this information (Fig. 2.5).

All of the interactions with the illness protocols (including the ones that lead to self-care) are recorded in the patient's record automatically. Therefore, I may see a patient for a check-up or a follow-up visit and I will notice in their record that they had what appeared to be a cold three weeks earlier and that they obtained information from the health education library. This information gives a more robust picture of my patient's complete health history. In addition, it often delights a patient when I inquire about an illness that they never directly discussed with me, my partners or our staff.

Another service available at home is the *Initial Health Assessment and Health Risk Appraisal*. This is an automated questionnaire that was designed using a number of well researched and useful questionnaires that have been in use in written form and, again, were reviewed by clinicians in a variety of specialties. Members are able to use their terminal at home to complete the health history and the health risk appraisal. The information from the health assessment is formatted into positive and negative findings. The health risk appraisal is organized to give a risk score and to highlight risk areas for the patient. If the patient has completed these questionnaires before they first come into the office, the patient or family and I are able to focus our time on what's important for that patient. In addition, health risk issues can be addressed in a very focused manner appropriate to that particular patient's needs.

We also offer *Health Highlights* to members at home. These include updates and information on current health topics. For instance, in the

```
===================>TES<===================

              Fever Treatment                   |
        Making Your Child Comfortable           |
                                                 |
Fever is a sign of illness or infection          |
and isn't necessarily bad. If your child         |
is uncomfortable, you can:                        |
                                                 |
*Dress your child in layers so he/she will       |
  cool off (but not get chilled)                 |
*Give plenty of fluids. At least 1/2 ounce       |
  every 30 minutes.                              |
*Get your child to rest. Most will limit         |
  their own activity.                            |
                                                 |
You can help by providing quiet activity         | <F1>Help
such as listening to music, reading stories      | <F2>Jump Back
or playing with a cuddly toy.                     | <F3>
                                                 | <F4>
                                                 | <F5>
                                                 | <F6>
                                                 | <F7>
Press <Enter> to continue                        | <F8>Main Menu
                                                 |
```

FIG. 2.5.   Health education information about fever treatment in triage and education system (TES).

summer the patients may be reminded to use sunscreen to prevent sunburns and they may be offered information about ticks and how to avoid them. In the late fall patients may be offered information about the impending flu season. We can also alert patients about high risk categories for complications of influenza in the winter. Those patients who are high risk are then urged to have flu shots and a variety of different flu shot sessions will be indicated.

Finally, a *suggestion box* comes with the home terminal. Members are able to send us requests or suggestions for health highlight topics, additional health education topics that they would be interested in, additional illness protocols that would be of value to them or suggestions about how to improve the system.

Information to the Home Pilot Study.

We conducted a pilot study of the system in 103 households (involving 256 people) who were randomly selected to have the system from our outpatient population. Our study lasted approximately six months and had the following findings:

- 95% of our members found that the system was easy or very easy to use in the various aspects that were rated.
- 85% of our members were satisfied or very satisfied with their use of the system.
- Over the study period, the participants increased their use of the system to decide when to call the doctor.
- The test group had 5% fewer visits in internal medicine during the 6-month study period compared with the rest of our patient population. This value was statistically significant.
- Members using the system developed increased confidence in self-care.

In summary, the home information pilot study has been successful and it appears that a system such as this is able to:

1. Provide valuable information to members in a useful format
2. Address many of their health care needs in a highly organized way that gives them great depth of detail
3. Document such illnesses and concerns, automatically updating members health records
4. Reduce the demand for unnecessary clinical visits.

Computerized Clinical Information System

The medical records for all of the patients of HCHP's Health Centers are computerized. Our center has been using a newer, more flexible system. Such systems are essential tools in the effort to effectively manage and coordinate the care of patients in managed care.

Primary care doctors and specialists in HCHP are all able to access our systems from various locations to be properly informed about a patient's history and to update their records. Staff in emergency rooms and hospitals and in our telecommunications centers can get up-to-date information from the systems as well.

Such an automated network is key to supporting our clinicians in their efforts to deliver high quality and efficient care. They can be well-informed about a patient's history, other clinician's findings, treatment plans, and orders. Appropriate interventions, testing, and treatment is facilitated. Unnecessary or redundant services (and costs) are avoided.

The key values of a clinical information system, such as we have used at HCHP's Burlington Center, includes:

- Instantaneous access to an automated medical record from multiple sites at any time, including physician's homes.
- Highly flexible searches for information
  - Selection and viewing of some (or all) of the information under a specific problem or diagnostic heading, i.e., *flowing* a problem (diagnosis) (Fig. 2.6).
  - *Flowing* of test results (Fig. 2.7).
  - *Flowing* of medication history (Fig. 2.8).
  - Access to multiple types of information simultaneously (Fig. 2.9)
- Automated laboratory test and x-ray ordering.
- Automated ordering of consultations
- Automated prescription ordering with simultaneous recording of prescription information in the note for that visit and in the patient's master medication list, and printing of the prescription in the pharmacy. (Patients have been thrilled that their medication is almost always ready for them to pick up when they arrive at the pharmacy since it was ordered through the system before they arrive.)

Efficient test results management is another key value. Actions that can be taken from the *Results Workbench* include: (a) automatic printing of a *normal* results letter listing the tests involved and including the values for tests patients often ask about (e.g., cholesterol or lead); (b) printing of a letter including the tests involved and text added by the clinician; (c) cre-

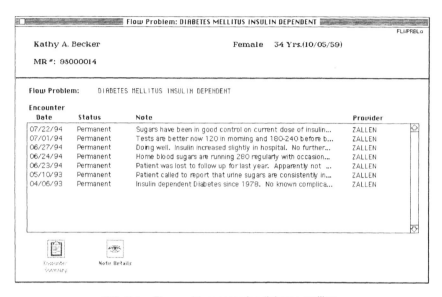

FIG. 2.6. Flow problem notes for diabetes mellitus.

Flow Problem: DIABETES MELLITUS INSULIN DEPENDENT

FLilPRBLa

Kathy A. Becker   Female   34 Yrs.(10/05/59)

MR *: 98000014

Flow Problem: DIABETES MELLITUS INSULIN DEPENDENT

Encounter

Date	Status	Note	Provider
07/22/94	Permanent	Sugars have been in good control on current dose of insulin...	ZALLEN
07/01/94	Permanent	Tests are better now 120 in morning and 180-240 before b...	ZALLEN
06/27/94	Permanent	Doing well. Insulin increased slightly in hospital. No further...	ZALLEN
06/24/94	Permanent	Home blood sugars are running 280 regularly with occasion...	ZALLEN
06/23/94	Permanent	Patient was lost to follow up for last year. Apparently not ...	ZALLEN
05/10/93	Permanent	Patient called to report that urine sugars are consistently in...	ZALLEN
04/06/93	Permanent	Insulin dependent Diabetes since 1978. No known complica...	ZALLEN

Encounter
Summary   Note Details

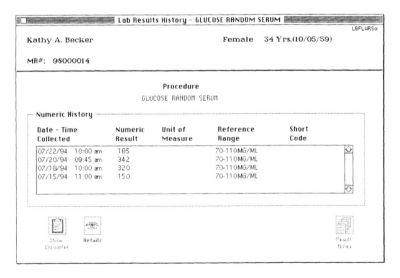

FIG. 2.7. Flow glucose (blood sugar) results.

Lab Results History - GLUCOSE RANDOM SERUM

LBFLWRSa

Kathy A. Becker   Female   34 Yrs.(10/05/59)

MR#: 98000014

Procedure
GLUCOSE RANDOM SERUM

Numeric History

Date - Time Collected		Numeric Result	Unit of Measure	Reference Range	Short Code
07/22/94	10:00 am	185		70-110MG/ML	
07/20/94	09:45 am	342		70-110MG/ML	
07/18/94	10:00 am	320		70-110MG/ML	
07/15/94	11:00 am	150		70-110MG/ML	

Show   Details   Result
Encounter   Notes

32

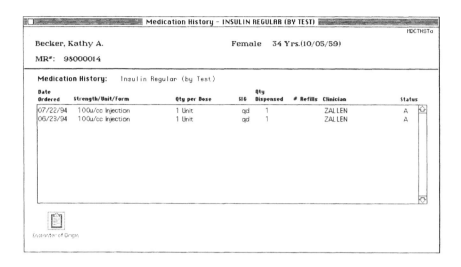

FIG. 2.8.   Flow insulin orders.

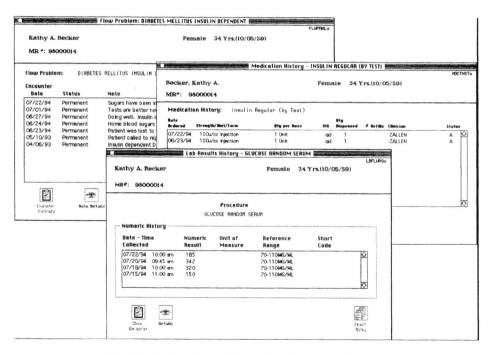

FIG. 2.9.   Multiple types of information viewed simultaneously.

ation of a telephone encounter to document a conversation, the plan for the patient, and to order further tests or prescriptions. Decision support is integrated into the automated medical record system, and to top it off—everything is legible.

In addition to our efforts with a clinical information system, many others have been working to use information technology to improve medical practice (Automated medical records, 1993; Computerized network, 1993; Council on scientific affairs, 1990; Grossman, 1994; Hennessey, Kanthak, Everett, & Lyons, 1993; Payne & Savarino, 1993; Schoenleber & Elias, 1993)

Teleradiology

Our center was also the test site for a system (R-Starr) to digitize our plain film x-rays and transmit the images over fiber-optic cable to Massachusetts General Hospital where the images were recreated for interpretation by appropriate radiologic specialists (Goldberg et al, 1993). Their findings were promptly faxed or telephoned back to us. This is also an example of how the new media can support coordinated, high quality care in a complex delivery system.

Improving Quality of Care

We are able to use the clinical information system to efficiently compare the outcomes for members with the same diagnosis, and similar presentations and test results, who are treated or managed in different ways. For example, we could compare the efficacy of treating asthma with one management approach versus another in a group of patients of similar age and with comparable histories, risk factors, and respiratory difficulty. The results of such comparisons would help us identify the approach that will be most successful and all of our clinicians could then adopt that method, and offer it to our members as standard procedure. Plan providers could share it with colleagues at other sites (Computer records, 1993). This enhances clinical quality, reduces unnecessary (and possibly detrimental) variation, and uses resources in the most effective manner leading to better care at less cost.

Further, if such systems were in widespread use, new disease patterns in a community, state or even the country could be more rapidly identified. For instance, the increase in unusual malignancies and immune deficiency in homosexual men in the early 1980s may have been identified earlier had automated clinical information systems been in place nationally (American Health Security Act of 1993, pp. 109–114, 136–140).

There has been a very strong effort at HCHP, and increasingly in all aspects of health care, to integrate the tools of quality management into the delivery of care. HCHP has been a major supporter and developer of the Health Plan Employer Data and Information Set (HEDIS). This is one of the first steps to define measures of quality which will be essential in efforts to improve health care delivery, service, and quality.

An automated clinical information system will be essential to producing such data and to further refine quality and value information. Many participants in HEDIS currently rely only on claims data. This will limit the ability to analyze outcomes, compare effectiveness of treatments, and to understand practice variation. The information derived from clinical information systems will contribute to a truly valuable health plan "report card" (American Health Security Act of 1993, pp. 99–101; Making health plans prove their worth, 1993).

Our automated clinical information system also allows us to generate clinical reminders. Several examples include:

- Patients identified in the system as having a risk factor for complications of the flu (e.g., asthma or age > 65 years), will be part of a report each fall that leads to printing of mailing labels for reminder postcards.
- Women overdue for a mammogram can be identified and contacted.
- Orders made for tests that have not yet been performed can be reviewed on screen or in a printed report and our staff can follow up with the patients to make sure they have the test we intended.

Confidentiality

Confidentiality of patient information in automated medical records has been raised as a concern. This issue has entered the health care reform debate (Gostin et al., 1993). In our use of an automated clinical information system, we have had no breaches of confidentiality. Our system requires a personal sign-on password and the level of access allowed varies by type of staff (e.g., clinician vs. medical assistant). To close (*sign*) an encounter requires a different personal password. Only clinicians can *sign* notes. Only a physician may prescribe a medication, of course, and they must use their personal *signature* password.

All interactions with the system are documented so that it will be clear who accessed the record, did what, and when. This automatic audit trail will be valuable in holding all users accountable for what they do with the system.

This feature of automated clinical information systems in general will be important not just for confidentiality and security, but also for prevention and detection of fraud.

In contrast to these levels of security, anyone who can get their hands on a *paper* medical record (or a copy of it) can breach confidentiality. Clearly, this will need to be debated further, but our experience suggests that automated systems can be very secure and can maintain confidentiality.

MANAGED CARE AND ADVANCED MEDIA
TECHNOLOGIES

The HCHP is currently experimenting with relatively basic new media technologies—a computerized information system and an interactive network. Although we have not yet experimented with the full range of new media, we are enthusiastic about the capacity of advanced media technologies to assist our providers. Members improve their decision-making capacity and clincians can further extend the decision support services to where and when health care decisions are made.

Advanced Media And Home Care

Using small, portable personal monitoring devices, data such as blood pressure, pulse, and glucose (sugar) could be transmitted from the patient's home to the doctor's office. This would be especially useful for patients who have stable but chronic conditions (such as hypertension or diabetes) and require routine monitoring. Further, audio transmission would aid in lung auscultation. Video transmission would allow for evaluation of skin color, rashes, assessment of degree of patient distress, and evaluation for infection of wounds, for instance.

These various monitoring devices could be controlled by a nurse or medical aide or even the patient alone. For instance, many patients know how to check their blood pressure; the device itself could be connected to a terminal or the patient could type in the values to be transmitted to the doctor's office. The patient, nurse, or health aide could place a stethoscope on the patient's chest for the doctor to listen to the digitized sounds. Video images of the chest could also be sent over the network. By adding interactive video teleconferencing capacity to home care the provider's "visit" can be personal and satisfying for the patient and the doctor.

Systems such as this in the home present a number of advantages for patients with obstacles to mobility. These include: the elderly, parents with infants and/or young children, any patient if there is inclement weather, patients with no transportation, patients who are physically challenged or have other handicapping conditions that may make it difficult for them to travel to the doctor's office.

For individuals and families in all these situations, the ability to have an evaluation by the doctor and a discussion with them without having to leave their home provides for high quality care with much greater convenience for the patient.

Further, as addressed in Dr. Deede's chapter (9), more interactive, multimedia triaging systems should be possible with the use of video and hypermedia . I would envision that patients would be able to browse through a virtual multimedia library and select the topics that they were interested in.

At present, most patient health education is directed by clinicians. In other words, a physician or other clinician hands a patient a brochure or refers them to a health education source for a particular topic. However, as technology improves and our access to such information (and the information itself) becomes more varied and interactive, patients will generate their own searches for such information far more often than clinicians do.

Support groups could be facilitated by the use of such systems, especially when video is a component. Patients would no longer have to travel to a central location in order to form a support group.

In addition, mental health groups could be connected through such a network and facilitated by a group leader even if all of the patients and the group leader are at their respective homes. Removing the obstacles of time, transportation and inconvenience may make it easier to enroll more patients in such valuable groups.

Expanded Managed Care Network

Today, HCHP's managed care network serves more than 500,000 members from New Hampshire to Rhode Island but the systems and information to facilitate care are only in a few locations (e.g., our offices).

I envision expanding our systems network to as many places health care decisions are made as possible. By bringing information and support tools to the point of decision making we will further enhance our cost effectiveness and quality of care by extending the intelligence of the medical system out to the point-of-service delivery.

Triage and education systems could be placed not just in patients' homes but also where they work by expanding the reach of a managed care network to the workplace. Employers who pay for their employees health care coverage have an incentive to be included in a managed care network. Employees who have real-time access to health information services may prevent or reduce the need for services and save the costs these unnecessary services incur. The "world's best phone message" could, with appropriate confidentiality protections , be triggered during the 8–10 hours of each day an individual spends at work if he or she could interact with their managed care organization from their desktop computer.

More universal access to such systems would improve access to and convenience of health care. Advanced systems employing interactive video or simpler ones such as our basic system could be available not only at home or at the workplace, but also at schools or even in public places such as in a closed booth at a shopping mall. In fact, such units might function like phone stations (perhaps they would be called *health stations* or *health booths*).

Expanding a managed care network and its computerized patient record to specialists can further contribute to the coordinated management of services. As described by Gorry et al (chapter 5), coordinated health care is greatly facilitated when experts can function as a team. The pharmacist, lab technician, radiologist, and other health and social service specialists could be entry points and/or contributors to a coordinated system of care.

As is discussed in Dr. Preston's chapter (4), such systems allow for physicians at different sites to consult about a patient. The primary physician, the specialist, and the patient are able to share video and audio information and other data and the patient could be at their home during this whole interaction.

Many in the health care reform debate have envisioned access to patients' medical histories from a variety of locations to enhance communication among the physicians caring for a patient, avoid redundant testing, improve quality of care, and to better serve the patient. An automated clinical information system network would make this vision a reality.

Automated systems should allow for better integration of practice guidelines into actual practice (American Health Security Act of 1993, pp. 109–114, 136-140; Payne & Savarino, 1993). Many physicians are aware of practice guidelines and many have participated in developing such guidelines. However, they are only rarely used in regular practice. This is due, at least in part, to the cumbersomeness of referring to such guidelines in a doctor's library or office while in the midst of a busy practice session. In addition, the guidelines themselves are designed or presented in a cumbersome fashion. Further, the information in the guidelines may become out of date or may conflict with information from another group that developed a guideline. All of these obstacles have led to minimal adoption of practice guidelines in a large segment of practice.

On-line access, the ability to modify and update guidelines quickly over a network, and integration into a patient's record through a clinical information system should facilitate the use of guidelines.

CONCLUSION

Systems such as I have discussed are clearly leading to an improved medical record and greater ease of consultation among clinicians. In addition, they provide for prompt availability and coordination of information from the home, workplace, physicians' offices, specialists' offices, laboratories, pharmacies, and hospitals. Further, there is improved organization and legibility in the medical record. Also, there is improved manipulation of and access to data in the record.

Member-centered systems provide a powerful health care reform tool for continuous quality improvement, clinical outcomes research, recognition of disease patterns across regions (potentially across the country) and easily available quality data integral to assessing and comparing health systems.

All of these valuable outcomes contribute to an improved quality of care for all patients.

In addition, the systems described will lead to much improved patient service by greater convenience and by avoiding unnecessary trips or hardships.

The uses of technology envisioned will gather the unserved and undeserved into the family of health. Groups isolated from access to care by geography, lack of transportation, or poverty will have access if these systems are guaranteed to all, analogous to a right to phone service. The brave new world of interactive systems in the home through cable and/or phone service will best be implemented by requiring a basic, low-cost, *lifeline* service available to all which would include access to healthcare systems such as discussed in this chapter. This would lead to greater democratization of health care.

Finally, the systems of the present and future hold the promise of greater personal empowerment of patients to obtain answers, guidance, and details about their health at their convenience. This will often be done without waiting for a mediator (practicing doctor) to obtain the information for them. When a clinician actually is involved they will be adding their particular skills and knowledge to what the patient has already been able to undertake for themselves. This technologically supported partnership should be a model for our future community.

APPENDIX A

Accidental poisoning	Bottlefeeding
Achilles tendon injuries	Breast self-exam
Acne	Breastfeeding techniques
Aerobic exercise	Bronchitis
AIDS	Bruxism
Allergic reactions	Burns
Angina	Caffeine
Ankle sprain or pain	Calcium
Anxiety	Car safety for children of all ages
Are you a problem user of alcohol and drugs?	Cardiovascular endurance
	Changes in diet (child)
Athlete's foot	Changes in diet (infant)
Barrier contraception methods	Changing your cholesterol level
Bicycle safety	Chicken pox
Bites and stings	Chills treatment
Boils	Chlamydia

Cholesterol and fat
Choosing the right contraception
Colds and flu (adult)
Colds and flu (child)
Colds and flu (infant)
Colic
Common breastfeeding problems
Common skin problems (itchy, dry skin)
Communication that works
Congestion (adult)
Congestion (child)
Congestion (infant)
Constipation (adult)
Constipation (child)
Constipation (infant)
Contact dermatitis
Coping with allergies
Corns and calluses
Coronary heart disease
Cough (adult)
Cough (child)
Cough (infant)
Cradle cap
Croup (ped)
Dandruff
Depression
Diaper rash
Diarrhea (adult)
Diarrhea (child)
Diarrhea (infant)
Dietary salt
Dysmenorrhea
Ear infections (adult)
Ear infections (child)
Ear infections (infant)
Ear wax (adult)
Ear wax (ped)
Eating for vitality
Elbow problems
Exercise and fitness
Exercise excuses and myths
Expressing and storing breastmilk
Fainting
Febrile seizures
Feeding or overfeeding?
Feet and toe problems
Fertility awareness

Fever (adult)
Fever (child)
Fever (infant)
First aid basics
Flouride
Food allergies (ped)
Foreign body in eyes, ear, nose
Gardnerella
Genital warts
Giardia (ped)
Gonorrhea
Good eating guidelines
Hands and finger problems
HCHP membership problems
HDL and LDL cholesterol
Health and well being at mid-life
Healthy sleep habits (ped)
Healthy snacks for kids
Hearing aids and assistive listening devices
Hearing loss
Heart attack
Heart disease risk factors
Helping your child feel safe
Helping your child sleep through the night
Hemorrhoids
Hepatitis
Herpes simplex
High blood pressure medications
Hip problems
Hodgkin's disease
Homemade baby food
Hormone replacement therapy
Hot flashes
How much sleep is enough?
How to limit the fat in your diet
How to read food labels
How to start your exercise program
How to take your temperature
How to tell if your child is having breathing problems (ped)
Indigestion
Insomnia
Iron in your diet
Is mealtime a battleground? (child)
Is mealtime a battleground? (infant)

Jock itch
Keeping balance in your life
Knee pain
Lactose intolerance
Laryngitis
Lead poisoning
Living with hot flashes
Low back pain
Lyme disease
Mammography
Managing hypertension
Menopause
Migraine headache
Mild bleeding
Minor wound care
Mouth blisters and sores (child)
Mouth blisters and sores (adult)
Neck pain
Night terrors
Nightmares
Nightwaking
Nose bleeds (ped)
Nursing twins
Nutrition while breastfeeding
Overuse injuries
Pap smear
Pediatric immunizations
Pelvic exam
Peptic ulcer
Physical activity readiness questionnaire
Plaque
Pneumonia
Poison ivy, oak, sumac
Poison proofing your home
Poisoning in adults
Poisoning in children
Precautions when taking aspirin or
    ibuprofen
Premenstrual syndrome (PMS)
Preventing choking
Preventing food poisoning
Preventing osteoporosis
Preventing tooth decay in children
Problem solving
Protein
Puberty explained for boys
Puberty explained for girls

Pubic lice
Quit for life (smoking cessation)
Relaxation
Safe food preparation
Season affective disorder (SAD)
Sex and the menopausal woman
Sexually transmitted diseases
Shin problems
Shoulder problems
Sinus headache
Sinus infection
Sleep hygiene
Social support networks
Sore nose (ped)
Sore throat (adult)
Sore throat (child)
Sore throat (infant)
Spitting up
Sports safety for children
Starting solid foods
Stomach flu (adult)
Stomach flu (child)
Stomach flu (infant)
Strains and sprains
Strength and flexibility
Strep throat (adult)
Strep throat (ped)
Stress management
Stroke
Sudden cardiac death
Sunburn
Surgical contraception
Syphilis
Taking your own blood pressure
Talking with your child about sex
Teething
Tension headache
Testicular self-exam
The truth about sugar
Time management
Tips for flying with children
Tips for the traveling baby
Tips for the traveling child
TMJ problems
Toxic shock syndrome
Trichomonas
TV and children

Urinary incontinence
Urinary tract infections
Using car seats
Using condoms
Vaginal spermicides
Vaginitis
Vitamins and minerals
Vomiting (adult)
Vomiting (child)
Vomiting (infant)
Warm-up exercises
Warts
Water
Weaning
Weight for me (weight loss program)
Well baby/child exam schedules

What is CPR?
What to do in case of poisoning
Wheezing (child)
Wheezing (infant)
When permanent teeth come in
Wrist problems
Yeast infections
Your child's safety: Birth to 6 months
Your child's safety: 7 to 12 months
Your child's safety: 1 to 2 years
Your child's safety: 2 to 4 years
Your child's safety: 5 years
Your child's safety: 6 to 7 years
Your child's safety: 8 to 9 years
Your child's safety: 10 years

REFERENCES

American Health Security Act of 1993. *Bureau of National Affairs*, 99–101, 109–114, 136–140.

Automated medical records: Leadership to expedite standards development. (1993, April). *U.S. General Accounting Office* (GAO/IMTEC-93-17).

Computer records can track physician performance. (1993, November 8). *American Medical News*, 26.

Computerized network of patient records not far off. (1993, July 26). *American Medical News*, 6.

Council on Scientific Affairs and Council on Long Range Planning and Development of the American Medical Association (1990). Medical informatics: An emerging medical discipline. *Journal of Medical Systems, 14,* 161.

Goldberg, M. A., Rosenthal, D. I., Chew, F. S., Blickman, J. G., Miller, S. W., & Mueller, P. R. (1993, February). A new high-resolution teleradiology system: Prospective study of diagnostic accuracy in 685 transmitted clinical cases. *Radiology. 186*(2); 429–434.

Gostin, L. O., Turek-Brezina, J., Powers, M., Kozloff, R., Faden, R., & Steinauer, D. D. (1993). Privacy and security of personal information in a new health care system. *JAMA, 270,* 2487.

Grossmann, J. H. (1994, January). Plugged-in medicine. *Technology Review.* 21–29.

Hennessey, J. M., Kanthak, M. H., Everett, E. R., & Lyons, J. P. (1993). A computerized radiology information system. *HMO Practice, 7,* 75.

Making health plans prove their worth. (1993, August 8). *The New York Times* [Business section]. p. 5.

Martin, A. R. (1993). The development of a computerized information system for clinicians. *HMO Practice, 7,* 56.

Payne, T. H., & Savarino, J. E. (1993). Computer "advice" aids in managing care of a population. *HMO Practice, 7,* 73.

Schoenleber, M. D., & Elias, S. (1993). The patient profile system: Group health's first iteration of the automated medical record. *HMO Practice, 7,* 67.

Wasson, J., Gaudette, C., Whaley, F., Sauvigne, A., Baribeau, P., & Welch, H. G. (1992). Telephone care as a substitute for routine clinic follow-up. *JAMA, 267,* 1788.

Zallen, B. G. (1993). Actual practice with InterPractice Systems: First experiences. *HMO Practice, 7,* 61.

Demand Management, Self-Care, and the New Media

Donald M. Vickery
Health Decisions International, LLC,
University of Colorado Health Sciences
Center

Health care reform has focused almost entirely on managing the supply of medical services, that is, *managed care* as it is currently practiced. Does management of the demand for medical services also deserve attention? If so, can the new media enhance demand management by facilitating self-care?

The belief that supply management alone can achieve at least one of the principal goals of health care reform, substantial cost containment, is open to question. For example, studies suggest that preadmission certification may reduce hospitalization by as little as 1% (Institute of Medicine, 1989) and utilization review may save only 1% of total medical care costs (Congressional Budget Office, 1992). Health maintenance organizations (HMOs) may reduce costs by as much as 10% (Wickizer, 1992), but such savings vary widely and may be due to selection by those with less demand for services (Liebowitz, Buchanan, & Mann, 1992). If there is a reasonable basis for the contention that demand management can contribute to cost containment while improving health and the quality of medical care, then it seems worthwhile to explore the question of how the new media might make effective demand management a reality for all Americans.

This chapter explores the relationships among demand management, self-care, the new media, and the goals of health care reform. It begins with an examination of the four components of demand: morbidity, perceived need, patient preference, and nonhealth motives, and their relevance to health, medical care costs, and the quality of medical care. This is followed by a discussion of self-care interventions as the tools of demand management and the evidence for the impact of these tools. After examining the technologies of self-care interventions, the effect of the new media on these interventions is discussed with emphasis on the potential for making support of individual decision making widely available.

DEMAND MANAGEMENT

Demand management may be defined as the support of individuals so that they can make rational health and medical decisions based on a consideration of the benefits and risks of the options available. Demand management is unabashedly rationalistic and information based. At the same time, it recognizes that decision making and actions are influenced by factors other than information; such as cognitive skills, social support, and cultural norms. Demand management addresses these issues as a part of its strategies to enable individuals to act on information.

Demand management is composed of four components: morbidity, perceived need, patient preference and nonhealth motives. The boundaries among these factors are not completely distinct; there is overlap and interaction among them as well as among the interventions that address these components. Nevertheless, it is useful to consider the components separately in order to construct a framework for exploring the strengths and weaknesses of demand management.

Morbidity

Morbidity is the most objective component of demand. Put simply, an illness or injury increases the probability that care will be sought. For some events, for example, a compound fracture, this probability may approach 100%.

The relationship between morbidity and utilization is not as strong as one might expect. Whereas morbidity consistently shows a positive relationship with utilization, available information suggests that it rarely accounts for more than 25% of the variance in utilization (Mechanic, 1979).

Similarly, the potential impact of prevention on utilization is not as great as might be expected and should be distinguished from its impact on health. For example, the potential benefits of prevention with respect to health are very large. Nearly two-thirds of all deaths and 70% of years of potential life lost (YPLL) are due to preventable causes. Yet only 30% of hospital days are due to these courses (Amler & Eddins, 1987).

Reductions in present or future levels of morbidity thus do not guarantee less demand for services, but they do hold substantial potential to do so. This fact increases the attractiveness of prevention as a cost-reduction strategy.

Health habits are strikingly associated with mortality, illness, and health status (Berkman & Breslow, 1983). Indeed, health habits not only correlate with illness and costs but are also predictive of future illness and costs. In one employee group, for example, certain health habits were found to predict medical claims costs during the three years after initial assessment. Smoking and high alcohol consumption produced $227 and $398, respectively, in excess medical costs per year (Yen, Edington, & Witting, 1991). Similarly, health risk appraisal data have been shown to be associated with medical costs (Leigh & Fries, 1992): no smoking, no excessive drinking, no excess body mass, exercise, and seat belt use were associated with $372 to $598 reduced direct cost per person per year in a group of retirees.

Morbidity due to infectious disease and accidents is an especially attractive target for reduction. Every year, more than 740 million infectious disease events occur, resulting in 200,000 deaths, 2 million years of life lost, 52 million hospital days and nearly 2 billion days lost from work, school, or other major activities. Annual medical care for these events costs an estimated $17 billion (Bennett, Holmberg, Rogers, & Solomon, 1987). Current preventive measures, if fully applied, could save an estimated $1.3 billion in direct costs, 56 million cases of infection, 3.2 million hospital days, 144 million disability days, 80,000 deaths and 1 million years of life lost annually (Bennett et al., 1987). Immunizations have already proven their effectiveness, of course, as is evident in the 99% reduced incidence of poliomyelitis, rubella, measles, and diphtheria (Centers for Disease Control, 1988).

Similarly, unintentional injuries each year cause more than 105,000 deaths, 2.7 million potential years of life lost, 48.8 million visits to emergency rooms, 27.7 million days of hospitalization, and 778.9 million days lost from major activities (Smith & Falk, 1987). An estimated 75% of motor vehicle injuries, 50% of home injuries, and 40% of occupational injuries could be prevented. These would presumably have related reductions in fatalities, medical care utilization, and costs (Smith & Falk, 1987). Reduction of alcohol use alone, through a broad-based, systematic approach, is estimated to reduce all types of injuries and associated deaths, medical care utilization, and costs by 25% (Smith & Falk, 1987).

It should be noted that secondary prevention, that is, screening, does not bear the same relationship to utilization as does primary prevention. Detection of asymptomatic disease increases utilization in the short term, whereas decreased utilization due to avoidance of complications may be delayed for many years or may not offset the costs of screening, follow-up, surveillance, and treatment of lifetime latent disease. The net effect may be to increase utilization of medical care (Russel, 1993). For example, lifetime screening of the elderly U.S. population for colon cancer has been estimated to cost $1.5 billion to $2.6 billion after savings in treatment costs are factored in (Wagner,

Herdman, & Wadhwa, 1991). The primary benefit of effective screening is not reduced utilization, but increased years of healthy life—44,000 to 61,000 years added in the case of colon cancer screening of the elderly.

Perceived Need

Perceived need is the primary basis upon which individuals make decisions to seek medical care. Although morbidity obviously plays a role in such decisions, a number of other factors also influence an individual's perception of need. These include knowledge of the risks and benefits of medical care, ability to assess the medical problem, perceived severity of the problem, ability to self-treat or self-manage the problem, and confidence in one's ability to manage the problem (self-efficacy). In turn, these variables are influenced by information, cultural norms, education, social support systems, comorbidity, and the attitudes of physicians and other providers.

The impact of perceived need has been demonstrated in a variety of ways. An individual who believes a particular symptom requires medical care will, upon experiencing the symptom, seek care more often than an individual experiencing the same symptom who does not share the belief (Tanner, Cockerham, & Spaeth, 1983). The decision to seek care is strongly influenced by the degree to which a symptom interferes with daily activities (Berkanovic, Telesky, & Reeder, 1981), as well as the presence of stressful life events combined with inadequate or distant social networks (Berkanovic et al., 1981; Counte & Glandon, 1991). People who perceive themselves to be in poor health use more medical care services than others (Connelly, Philbrick, Smith, & Wymer, 1989), even when physicians judge them to be physically healthy (Connelly, Smith, Philbrick, & Kaiser, 1991). Overall, morbidity has been estimated to account for 12% of explained variance in decisions to seek care, whereas personal beliefs and social network account for 42% (Berkanovic et al., 1981).

The decision to seek care cannot be predicted on the objective basis of severity. Among hearing-impaired individuals, for example, one study found no relationship between severity of impairment and increased use of medical services, suggesting that additional utilization may reflect social isolation experienced by those individuals (Kurz, Haddock, Van Winkle, & Wang, 1991).

Assuming that individuals perceive a need for services when they make outpatient visits, enter the hospital or undergo procedures, overall use and costs of services are greatly affected by varying perceptions of need. This is illustrated by studies that use panels of physicians to judge "appropriateness" of medical services. Table 3.1 displays some of the best known estimates (Bernstein et al., 1993; Chassin & Kosecoff, 1989; Chassin, Kosecoff,

TABLE 3.1
Inappropriate or Equivocal Medical Care

Service	Inappropriate or Equivocal
Coronary Artery Bypass Graft (Winslow et al., 1991)	44%
Coronary Artery Bypass Graft (Leape et al., 1993)	9%
Carotid Endarterectomy (Winslow et al., 1991)	64%
Coronary Angiography (Graboys et al., 1992)	94%
Coronary Angiography (Chassin et al., 1987)	24%
Coronary Angiography (Bernstein et al., 1993)	24%
Coronary Angioplasty (Hilborne et al., 1993)	42%
Tonsillectomy (Roos, 1979)	86%
Hospital Days (Siu et al., 1986)	35%
Gastrointestinal Endoscopy (Chassin et al., 1989)	28%

Solomon, & Brook, 1987; Graboys, Biegelsen, Lampert, Blatt, & Lown, 1992; Hilborne et al., 1993; Leape et al., 1993; Roos, 1979; Siu et al., 1986; Winslow, Kosecoff, Chassin, Kanouse & Brook, 1988; Winslow et al., 1991) of equivocal or inappropriate care. According to this methodology, the bulk of evidence suggests that one-third or more of medical care is likely to be judged inappropriate or equivocal by physician panels, yet the patients and their doctors must have perceived a need for it. These studies do not establish what is right or wrong in medical care, but they conclusively demonstrate widely varying perceptions of need and their impact on the use of medical services. If the patients and the doctors had the same perceived need as the panels, utilization would have been dramatically reduced.

Patient Preference

Patient preference is the critical factor in choosing among available diagnostic and treatment approaches when patients and physicians share decision-making appropriately. The basic principle of appropriate decision-making is relatively straightforward and appeals to common sense: Choose the option that offers the greatest benefit for the least risk and cost, an approach that may best be termed *economic appropriateness* because it includes cost. Approaches that exclude cost and choose options offering the greatest benefit for the least risk may be termed *medical appropriateness*; a convincing argument can be made, however, that cost considerations are unavoidable (Eddy, 1990).

Several factors hinder assessment of economic appropriateness in medical decision making. First, medical science often provides surprisingly little information with respect to the probable benefits or hazards of a particular option (Eddy et al., 1987). Second, the costs of options may be ignored in the decision-making process because traditional insurance plans did much to remove cost as a conscious consideration, and because physician training does not focus on cost issues, leaving many physicians with strong biases against considering costs when making diagnostic and management decisions. Third and most important, patients are rarely encouraged to weigh the benefits against the risks and costs of the available options, informed consent requirements notwithstanding. More often, a particular treatment or approach is proposed as the *best choice*, a reflection of the doctor's judgment, not the patient's.

In most states, the legal standard of informed consent is based on a model of *informed choice* (E. J. Emanuel & L. L. Emanuel, 1992). According to this model, physicians explain treatment/service options and the probabilities of their associated outcomes, but patients determine the utility of outcomes and select an option based on their preference. Informed choice has the advantage of making clear that the physician is responsible for describing the risks and benefits of all the options available, even though some studies indicate that physicians selectively omit this information to obtain informed consent (Howard & DeMets, 1981; Wu & Pearlman, 1988). It also clarifies the difference between benefit/risk probabilities associated with an outcome and the values (*utilities* in economic terms) that an individual patient places on those outcomes. Finally, it delineates the unequivocal role of the patient as an active decision maker. Some medical self-care programs may have had limited effectiveness because of confusion with respect to the patient's role in shared decision making with physicians (Cassileth, Zupkis, Sutton-Smith, & March, 1980; Greenfield, Kaplan, & Ware, 1985; Greenfield, Kaplan, Ware, Yano, & Frank, 1988; Kaplan, Greenfield, & Ware, 1985; Korsch & Negrete, 1972; Roter, 1977). The informed choice model provides the clarity needed for progress in this area.

Most important, it indicates that valuation of risks and benefits necessary for appropriate decision making cannot be accomplished without patient participation. In short, there cannot be appropriate care without the expression of patient preference.

A growing share of research attention is being devoted to evaluating benefits and risks of various diagnostic and therapeutic options for some situations in clinical medicine (Institute of Medicine, 1990), often with the goal of producing guidelines or protocols for clinical decision making. As this research progresses, it is increasingly clear that patient preference must be an explicit part of these protocols, because the comparison of various benefits and risks usually presents an "apples and oranges" dilemma involving trade-offs. A good example is the choice between surgical and nonsurgical therapy for patients with benign prostatic hypertrophy. Those who choose surgery are more than twice as likely to experience improved symptoms but also suffer potentially severe side effects in one of four cases. When 284 HMO patients in Denver viewed these odds in a videodisc produced at the Dartmouth Medical School, only 30 chose surgery while 254 postponed it (Dentzer, 1991), a 50% decline in the expected surgery rate for this condition. The result suggests that previous patients were not fully informed of risks and benefits and that physician preferences were substituted for patient preferences. The result also confirms other studies suggesting that patients are more risk-averse than physicians in most situations (Wennberg, 1990). More importantly, it demonstrates how patient preference can dramatically alter the practice of medicine and surgery, help lower the costs of medical care, and help make appropriate care also affordable care.

Some recent clinical guidelines clearly incorporate the role of patient preference. For example, cataract operations are recommended when poor vision interferes with life to the point that the *patient* believes that the potential benefit outweighs the risks of surgery (infection, loss of the eye, etc.) (U.S. Dept. of Health and Human Services, 1993). This benefit is valued differently by someone whose lifestyle requires a relatively high level of visual acuity, for driving or work, than by someone who is sedentary, and does not drive. Surgery is not done simply because the probability of improved vision is greater than the probability of a significant complication.

Care for terminal illnesses presents another opportunity for patient preference to influence utilization. Medical care expenditures increase as much as 80-fold in the last year of life (Roos, Shapiro, & Tate, 1989). The value of terminal care to the individual, especially the use of intensive care units, has been extensively discussed (Bone, 1993; Greco, Schulman, Lavizzo-Mourey, Hansen-Flaschen, 1991; Podrid, 1989). Currently, patient preference may be routinely ignored in this situation because, although most patients want advance directives and want life-sustaining treatment discontinued if they become incompetent and have a poor prognosis (Emanuel, Barry, Stoeckle,

Ettelson, & Emanuel, 1991), only 9% of Americans had written advance directives in 1987 (Steiber, 1987) and physicians often do not know of these directives when they do exist (Brennan, 1988). The routine use of advance directives such as living wills could have a dramatic impact on medical care utilization and cost.

Nonhealth Motives

Most physicians, health plan administrators, and employers recognize that individuals may use medical care for reasons not related to their health, such as qualifying for sick leave, disability, or worker's compensation benefits. This phenomenon sometimes has been referenced through the label *moral hazard* (Donaldson & Gerard, 1989), an economic term that many find awkward; *nonhealth motives* is proposed as an alternative.

It appears that nonhealth motives may be accentuated in certain situations. According to H. H. Gardner, M.D. (written communication, November, 1992), medical claims are increased when disability benefits are available. Utilization has been inversely correlated with the amount of the patient's copayment fee (with little or no impact on patient health as copayment fees increase) (Brook, 1991), a phenomenon that may be due at least partially to the effect on nonhealth motives.

Many employers believe that nonhealth motives increase their medical care costs. Unfortunately, there is little quantitative information on the overall impact of nonhealth motives on the use of medical care or on the impact of interventions, such as specialized case management, directed at nonhealth motives.

SELF-CARE AND DEMAND MANAGEMENT

Self-care (the actions that individuals take with respect to health and medical care) is central to demand management, simply because individual knowledge and behavior powerfully affect all components of demand. The practice of demand management, then, is largely the use of interventions support of self-care.

Self-care devoted to health, that is, health self-care, may be equated with prevention and directly affects morbidity. In contrast, medical self-care's relationship to demand is complex (Table 3.2), involving relatively distinct activities, each of which may affect multiple components of demand. The principal activities—self-management, informed choice, self-help, life management, screening, and provider information—and their effects on demand are discussed in the following sections (Lee, 1993).

Self-management, or the ability to manage one's own medical problems,

TABLE 3.2
Relationship of Demand Components to Self-Care Interventions

Patient Preference	Self-Care Intervention Demand Component		
	Non-Health Motives	Morbidity	Perceived Need
Health Self-Care			
	Risk Reduction (Lifestyle, Immunization, Safety)	X	
Medical Self-Care			
	Self-Management	X	X
	Informed Choice	X	XX
	Self-Help	X	X
	Life Management		XX
	Screening	X	
	Provider Information		X

affects both morbidity and perceived need. Accurate self-assessment may prevent progression or complications, whereas inaccurate assessment may produce perceived need at substantial variance from actual need. Self-assessment is usually coupled with self-treatment in self-management programs, especially for minor illnesses. However, self-management programs may be almost entirely directed at support of self-treatment of chronic diseases, such as arthritis, for which the fundamental assessment is unlikely to change or changes very slowly.

Support for the exercise of informed choice profoundly affects patient preference but may also affect perceived need, because an accurate assessment of options may alter the perception of need for medical care. By improving the appropriateness of medical care, morbidity is decreased. Informed choice interventions can also manage nonhealth motives. Individuals for whom it may be an issue are identifiable through excessive benefit use, and are addressed from the standpoint that they may not be managing problems optimally and could benefit from support in decision making. Such a strategy offers the advantage of approaching individuals on the basis of self-interest rather than as a policeman for benefits, increasing the likelihood of behavior change.

Self-help groups are organized and conducted by lay individuals who are dealing with a medical problem. Enhancement of self-efficacy, an individual's belief that he or she can effectively deal with a particular problem, may be the primary method by which self-help groups such as those for arthritis, diabetes, and alcoholism improve health status and reduce perceived need for medical care. However, self-help groups may also include self-management interventions.

Life management programs help individuals clarify goals and intentions regarding the highly charged issues of medical care when the people are dying and/or incapable of exercising informed choice. Such programs emphasize the use of instruments such as living wills and durable power of attorney.

Finally, provider information regarding specialty, fees, facilities, hours, location, and patient education programs may influence patient preference for a particular provider or site of care and how a provider's services are used.

IMPACT OF SELF-CARE INTERVENTIONS

Health self-care interventions have been extensively evaluated in the workplace (Pelletier, 1991), where they often emphasize health risk appraisals, individual counseling, group approaches, communications, or a combination of these approaches. The impact of selected programs (Bly, Jones, & Richardson, 1986; Bowne, Russell, Morgan, Optenberg, & Clarke, 1984; Centers for Disease Control, 1988; Gibbs, Muluaney, Henes, & Reed, 1985; Kahane, 1986; Mueller et al., 1988; Shepard, Corey, Ruezland, & Cox, 1982) on medical care costs is given in Table 3.3.

Medical self-care interventions use the same approaches as health self-care, but are more likely to emphasize communications technology than

TABLE 3.3
Effects of Selected Health Self-Care Programs

Lifestyle

Reduced medical claims by $360; benefit/cost of 3:1 (Bowne et al., 1984)
Reduced medical claims by $109 (Gibbs et al., 1985)
Reduced medical claims $32 (Bly et al., 1986)
Reduced medical claims by $84 (Shepard et al., 1982)

Immunization

Decreased (99%) incidence of: (Centers for Disease Control, 1988)

 poliomyelitis
 rubella
 measles
 diptheria

Safety

Decreased (66%) hospital admissions and charges for automobile accidents with seat belt use (Mueller et al., 1988)
Decreased (67%) serious injuries estimated for automobile accidents with child safety seats (Kahane, 1986)

TABLE 3.4
Effects of Selected Medical Self-Care Interventions

Intervention Category Effects

Self-Management

Decreased (17%) physician visits, decreased (35%) visits for minor illnesses (Vickery et al., 1983).
Decreased (75%) readmission rate and decreased (54%) hospital days for asthmatics (Mayo et al., 1990).
Decreased (22%) postoperative hospital stay with patient-controlled analgesia (White, 1988).
Decreased (69%) hospital admissions by diabetics (Miller et al., 1972).
Decreased (25%) specialist referrals and decreased (11%) per visit costs (Kemper, 1982).
Decreased (89%) hospital days, decreased (76%) outpatient visits, decreased (45%) health care costs for hemophiliacs (Levine et al., 1973).

Informed Choice

Decreased (17%) repeat caesarian sections (Gardner et al., 1990).
Decreased (2.7 days, 1 day) hospital stays for surgical patients (Egbert et al., 1964, Wilson, 1981).
Decreased (5.5–14%) inpatient claims costs (OPTIS Analysis Service Reports, 1992).
Decreased (50%) surgery for benign prostatic hypertrophy (Dentzer, 1991).

Self-Help

Decreased (43%) physician visits, decreased pain in arthritics (Lorig et al., 1993).

Life Management

Most (93%) patients want advance directives and most (71%) want life-sustaining treatment discontinued if they were to become incompetent and have a poor prognosis (Emanuel et al., 1991).

Screening

Increased medical costs estimated for hypertension (Russel, 1993).
Increased medical costs estimated for colon cancer (Wagner et al., 1991).

Provider Information

Decreased (24%) unscheduled office visits and increased satisfaction (Stirewalt et al., 1982).

face-to-face approaches. Effects of selected medical self-care interventions (Dentzer, 1991; Egbert, Battit, Welch, & Bartlett, 1964; Emanuel et al., 1991; Gardner & Sneiderman, 1990; Kemper, 1982; Levine & Britten, 1973; Lorig, Mazonson & Holman, 1993; Mayo, Richman, & Harris, 1990; Miller & Goldstein, 1972; OPTIS® Analysis Service Reports, 1992; Russel, 1993; Stirewalt, Linn, Gogoy, Knopka, & Linn, 1982; Vickery et al., 1983; Wagner et al, 1991; White, 1988; Wilson, 1981) are given in Table 3.4.

THE TECHNOLOGY OF DEMAND MANAGEMENT

The self-care interventions cited in Tables 3.3 and 3.4 are becoming widely available, and multiple interventions have been offered in combination as complementary services or products. For example, health risk appraisals, written and audiovisual media, telephone counseling services, and community resources have been combined to provide support for risk reduction, self-management of acute and chronic disease, informed choice regarding selection of diagnostic and treatment options, formation of self-help groups, tools for life management (such as advance directives), and provider information.

The communications technologies used for these interventions are not advanced: Written communications, the telephone, didactic classroom teaching, and individual instruction account for virtually all of the effective interventions with the notable exception of Wennberg's interactive video disk. Such low-tech approaches have the advantages of being practical and available at a reasonable cost.

Yet most of these interventions are not in widespread use, a fact that highlights their limitations. Didactic teaching, whether group or individual, is limited almost exclusively to chronic problems in which there is a continuing and sustained interest in one disease or process. Even when such interest exists, the barriers to bringing instructors and students together on a continuing basis are often insurmountable. The result is that self-care learning for asthmatics, diabetics, hemophiliacs, and others with conditions requiring long-term management, is still largely a hit-or-miss proposition. Much the same applies to self-help groups for problems such as arthritis.

Mass media can reach virtually all individuals, of course. In the form of reference materials, they have the advantage of being available when the individual decides that assistance is necessary, an important issue for helping with virtually all acute problems. On the other hand, such media are not interactive, personalized or capable of rendering social support. These characteristics are relatively unimportant when the issue is minor, but are prerequisites for dealing effectively with such major decisions as those concerning surgery, hospitalization, or chemotherapy. The telephone can meet these requirements in many instances. Further, it is widely available and individuals are willing to use it provided that they understand the purposes of the service and trust its confidentiality. It can be combined with written or audiovisual materials to overcome some of the inherent limitations of audio-only communication, but it does so at a cost in resources and in response times when multimedia responses are required. This complex approach may still appear fragmented to the user and coordinated use of multiple interventions for each and every user remains a challenge.

SELF-CARE AND THE NEW MEDIA

The new media can be expected to have substantial impact on self-care interventions in two areas. First, certain of the characteristics of the new media—interactivity, personalization, dimension, agency, and selectivity—directly address the limitations of the technologies currently in use. For example, the new media can accommodate the demands of users for self-care information that is immediately available, personalized to their issue, and presented in a multimedia approach that suits their learning and decision style while allowing them to control the management process.

Second, the new media's accessibility, combined with the attributes mentioned above, may finally eliminate the limitations on widespread use of self-care interventions, especially those that have in the past called for didactic teaching on an individual or group basis. It seems likely that any individual with a chronic problem could have access to the kind of learning experience that heretofore has been available only when inspired teachers and students have overcome the barriers to meeting on a continuing basis.

The new media will face their greatest challenge in attempting to meet the needs of individuals for emotional and social support in dealing with major health problems and decisions. This has been the forte of self-help groups and telephone-based decision support services. It may well be that the greatest gain will come from the intelligent integration of the new media with these human components rather than attempting to replace such interactions with simulated dialogue via computer programs. For example, the logic, displays, and vignettes of current interactive videodisc programs dealing with medical decisions could be integrated into encounters between clients and counselors that take place via broadband (video and voice) communications.

DEMAND MANAGEMENT AND THE NEW MEDIA

Perhaps the greatest impact of the new media will not be on enhancement of specific self-care interventions, but rather on the creation of a comprehensive system of demand management. The goal of this system will be the coordinated delivery of all the demand management (self-care) interventions for which there is acceptable cost effectiveness and/or benefit cost information. The hub of this system will be a health decision support service (HDSS) capable of providing both the information and psychosocial support necessary for rational decisions with regard to prevention, self-management, and participation in shared decision making with medical care providers. The HDSS will be linked to community resources and medical care systems as well as its clients so that support for self-care is coordinated, comprehensive, and convenient. Today's forerunners of this HDSS rely primarily on telephonic communications, but the HDSS of the future can be expected to make maximum use of advances in communication technology, for example, the information superhighway.

Central to such a system will be the development of a sophisticated, multipurpose information system. The development of a new structure for personal health and medical data—a true health record—will support interactions among individuals, the demand management system, and the medical care system. The record will contain health risk data, demographics, psychosocial information, and patient-derived process and outcome data from encounters with the medical care system. It may also contain abstracted medical, workers' compensation, and disability claim information. Given the legal and practical obstacles to integrating the health record with current medical records, it is likely that the health record will be separate from the medical record. This will continue to be in the province of the medical care system, but may exchange data with the medical record under certain circumstances.

In addition to supporting individuals, the system will support management and evaluation functions, including the integrated analysis of multiple data sets (medical claims, workers' compensation claims, disability claims, psychosocial data, demographics, health risk data, and outcomes data). These will guide the development and implementation of interventions, match interventions to individuals and subgroups, and evaluate the performance of these interventions. Patient-derived process and outcomes measures will be available for evaluating medical care systems.

Together, these coordinating and evaluating functions will enable consumers to understand, select, and evaluate medical care services in a way not yet possible because they lack access to such comprehensive services. Such a change helps to restore balance and equity in the relationship between consumers and providers.

CONCLUSION

Although demand management is a new concept, there is substantial evidence for the effectiveness, cost-effectiveness, and favorable benefit–cost ratios of the self-care interventions that are its tools. These tools are being coalesced into increasingly sophisticated and comprehensive programs of substantial power, but a comprehensive system of demand management has yet to be implemented. The new media can play a major role in overcoming the barriers to creation of such a system by enhancing the effectiveness of individual interventions, increasing the availability of the interventions, and, most importantly, providing a sophisticated, multipurpose information system based on a new concept of the individual health record.

Effective demand management not only has a favorable impact on medical costs, it is essential to improving both health and the quality of medical care, a claim that supply management alone cannot make. Thus attention to the demand for medical services is the key to attaining the goals of health care reform.

REFERENCES

Amler, R. W., & Eddins, D. L. (1987). Cross-sectional analysis: Precursors of premature death in the United States. In R. W. Amler & H. B. Dull (Eds.), *Closing the gap: The burden of unnecessary illness* (pp. 181–187). New York: Oxford University Press.

Bennett, J. V., Holmberg, S. D., Rogers, M. F., Solomon, S. L. (1987). Infectious and parasitic diseases. In R. W. Amler & H. B. Dull (Eds.), *Closing the gap: The burden of unnecessary illness* (pp. 102–114). New York: Oxford University Press.

Berkanovic, E., Telesky, C., & Reeder, S. (1981). Structural and social psychological factors in the decision to seek medical care for symptoms. *Medical Care, 19,* 693–709.

Berkman, L. F., & Breslow, L. (1983). *Health and ways of living.* Oxford, UK: Oxford University Press.

Bernstein, S. J., Hilborne, L. H., Leape, L. L., Fiske, M. E., Park, R. E., Kamberg, C. J., & Brook, R. H. (1993). The appropriateness of use of coronary angiography in New York State. *Journal of the American Medical Association, 269,* 766–769

Bly, J., Jones, R., & Richardson, J. (1986). Impact of worksite health promotion on health care costs and utilization: Evaluation of Johnson & Johnson's Live for Life Program. *Journal of the American Medical Association, 256,* 3235–3240.

Bone, R. C. (1993). Concepts in emergency and critical care: Intensive care, survival, and expense of treating critically ill cancer patients. *Journal of the American Medical Association, 269,* 783–786.

Bowne, D., Russell, M., Morgan, J., Optenberg, S., & Clarke, A. E. (1984). Reduced disability and health care costs in an industrial fitness program. *Journal of Occupational Medicine, 26,* 809–816.

Brennan, T. A. (1988). Ethics committees and decisions to limit care: The experience at the Massachusetts General Hospital. *Journal of the American Medical Association, 260,* 803–807.

Brook, R. H. (1991). Health, health insurance, and the uninsured. *Journal of the American Medical Association, 265,* 2998–3002.

Cassileth, B. R., Zupkis, R. V., Sutton-Smith, K., & March, V. (1980). Informed consent: Why are its goals imperfectly realized? *New England Journal of Medicine, 302,* 896–900.

Centers for Disease Control. (1988). Summary of notifiable diseases, United States, 1987. *Morbidity and Mortality Weekly Report, 36,* 1–59.

Chassin, M. R., & Kosecoff, J. (1989). *The appropriateness of selected medical and surgical procedures: Relationship to geographic variations.* Ann Arbor, MI: Health Administration Press.

Chassin, M. R., Kosecoff, J., Solomon, D. H., & Brook, R. H. (1987). How coronary angiography is used: Clinical determinants of appropriateness. *Journal of the American Medical Association, 258,* 2543–2547.

Congressional Budget Office. (1992). *The potential impact of certain forms of managed care on health care expenditures.* Washington, DC: Author.

Connelly, J. E., Philbrick, J. T., Smith, G. R., & Wymer, A. (1989). Health perceptions of primary care patients and the influence on health care utilization. *Medical Care, 27*(suppl.), S99–S109.

Counte, M. A., & Glandon, G. L. (1991). A panel study of life stress, social support, and the health services utilization of older persons. *Medical Care, 29,* 348–361.

Dentzer, S. (1991). The allegory of the conveyor belt. *Dartmouth Alumni Magazine, 5,* 23–25.

Donaldson, C., & Gerard, K. (1989). Countering moral hazard in public and private health care systems: A review of recent experience. *Journal of Social Policy, 18,* 235–251.

Eddy, D. M. (1990). Clinical decision making: From theory to practice. What do we care about costs? *Journal of the American Medical Association, 264,* 1161–1170.

Eddy, D. M., Nugent, F. W., Eddy, J. F., Coller, J., Gilbertsen, V., Gottlieb, L. S., Rice, R., Sherlock, P., & Winawer, S. (1987). Screening for colorectal cancer in a high-risk population: Results of a mathematical model. *Gastroenterology, 92,* 682–692.

Egbert, L. D., Battit, G. E., Welch, C. E., & Bartlett, M. K. (1964). Reduction of postoperative pain by encouragement and instruction of patients: A study of doctor-patient rapport. *New England Journal of Medicine, 270,* 825–827.

Emanuel, L. L., Barry, M. J., Stoeckle, J. D., Ettelson, L. M., & Emanuel, E. J. (1991). Advance directives for medical care: A case for greater use. *New England Journal of Medicine, 324,* 889–895.

Emanuel, E. J., & Emanuel, L. L. (1992). Four models of the physician-patient relationship. *Journal of the American Medical Association, 267,* 2221–2226.

Gardner, H. H., & Sneiderman, C. A. (1990). Ensuring value by supporting consumer decision making. *Journal of Occupational Medicine, 32,* 1223–1228.

Gibbs, J., Mulvaney, D., Henes, C., & Reed, R. W. (1985). Worksite health promotion: Five-year trend in employee health care costs. *Journal of Occupational Medicine, 27,* 826–830.

Graboys, T. B., Biegelsen, B., Lampert, S., Blatt, C. M., & Lown, B. (1992). Results of a second-opinion trial among patients recommended for coronary angiography. *Journal of the American Medical Association, 268,* 2537–2540.

Greco, P. J., Schulman, K. A., Lavizzo-Mourey, R., & Hansen-Flaschen, J. (1991). The patient self-determination act and the future of advance directives. *Annals of Internal Medicine, 115,* 639–643.

Greenfield, S., Kaplan, S. H., & Ware, J. E., Jr. (1985). Expanding patient involvement in medical care: Effects on patient outcomes. *Annals of Internal Medicine, 102,* 520–528.

Greenfield, S., Kaplan, S. H., Ware, J. E., Jr., Yano, E. M., & Frank, H. J. (1988). Patient participation in medical care: Effects on blood sugar control and quality of life in diabetes. *Journal of General Internal Medicine, 3,* 448–457.

Hilborne, L. H., Leape, L. L., Bernstein, S. J., Park, R. E., Fiske, M. E., Kamberg, C. J., Roth, C. P., & Brook, R. H. (1993). The appropriateness of use of percutaneous transluminal coronary angioplasty in New York State. *Journal of the American Medical Association, 269,* 761–765.

Howard, J. M., & DeMets, D. (1981). How informed is informed consent? The BHAT experience. *Controlled Clinical Trials, 2,* 287–303.

Institute of Medicine. (1989). *Controlling costs and changing patient care? The role of utilization management.* Washington, DC: National Academy Press.

Institute of Medicine. (1990). *Clinical Practice Guidelines.* Washington, DC: National Academy Press.

Kahane, C. J. (1986). *An evaluation of child passenger safety. The effectiveness and benefits of safety seats (summary).* DOT publication no. (DOT HS) 806–889. Washington, DC: National Highway Traffic Safety Administration.

Kaplan, S. H., Greenfield, S., & Ware, J. E., Jr. (1985, April). *Expanded patient involvement in medical care: Effects on blood pressure.* Paper presented to National Conference on High Blood Pressure Control, Chicago, IL.

Kemper, D. W. (1982). Self-care education: impact on HMO costs. *Medical Care, 20,* 710–718.

Korsch, B. M., & Negrete, V. F. (1972). Doctor-patient communication. *Scientific American, 227,* 66–74.

Kurz, R. S., Haddock, C., Van Winkle, D. L., & Wang, G. (1991). The effects of hearing impairment on health services utilization. *Medical Care, 29,* 878–889.

Leape, L. L., Hilborne, L. H., Park, R. E., Bernstein, S. J., Kamberg, C. J., Sherwood, M., & Brook, R. H. (1993). The appropriateness of use of coronary artery bypass graft surgery in New York State. *Journal of the American Medical Association, 269,* 753–760.

Lee, J. M. (1993). Screening and informed consent. *New England Journal of Medicine, 328,* 438–440.

Leigh, J. P., & Fries, J. F. (1992). Health habits, health care use and costs in a sample of retirees. *Inquiry, 29,* 44–54.

Levine, P. H., & Britten, A. F. H. (1973). Supervised patient-management of hemophilia: A study of 45 patients with hemophilia A and B. *Annals of Internal Medicine, 78,* 195–201.

Liebowitz, A., Buchanan, J. L., & Mann, J. (1992). A randomized trial to evaluate the effectiveness of a Medicaid HMO. *Journal of Health Economics, 11*(3), 235–257.

Lorig, K. R., Mazonson, P. D., & Holman, H. R. (1993). Evidence suggesting that health education for self-management in patients with chronic arthritis has sustained health benefits while reducing health care costs. *Arthritis and Rheumatism, 36,* 439–446.

Mayo, P. H., Richman, J., & Harris, H. W. (1990). Results of a program to reduce admissions for adult asthma. *Annals of Internal Medicine, 112,* 864–871.

Mechanic, D. (1979). Correlates of physician utilization: Why do major multivariate studies of physician utilization find trivial psychosocial and organizational effects? *Journal of Health and Social Behavior, 20*(12), 387–396.

Miller, L. V., & Goldstein, J. (1972). More efficient care of diabetic patients in a county hospital setting. *New England Journal of Medicine, 286,* 1388–1391.

Mueller, O. E., Turnbull, T. L., Dunne, M., Barrett, J. A., Langenberg, P., & Orsay,

C. P. (1988). Efficacy of mandatory seat belt use legislation. *Journal of the American Medical Association, 260,* 3593–3597.

OPTIS® Analysis Service Reports. (1992). Cheyenne, WY: Options and Choices, Inc.

Pelletier, K. R. (Ed.). (1991). A review and analysis of the health and cost-effective outcome studies of comprehensive health promotion and disease prevention programs. *American Journal of Health Promotion, 5,* 311–313.

Podrid, P. J. (1989). Resuscitation in the elderly: A blessing or a curse? *Annals of Internal Medicine, 111,* 193–195.

Roos, N. P. (1979). Who should do the surgery? Tonsillectomy-adenoidectomy in one Canadian province. *Inquiry, 16,* 73–83.

Roos, N. P., Shapiro, E., & Tate, R. (1989). Does a small minority of elderly account for a majority of health care expenditures? A sixteen-year perspective. *Milbank Quarterly, 67,* 347–369.

Roter D, (1977). Patient participation in patient-provider interactions: The effects of patient question-asking on the quality of interactions, satisfaction and compliance. *Health Education Monographs, 5,* 281–314.

Russel, L. B. (1993). The role of prevention in health report. (Editorial) *New England Journal of Medicine, 329*(5), 352–354.

Shepard, R., Corey, P., Ruezland, P., & Cox, M. (1982). The influence of an employee fitness program and lifestyle modification program upon medical care costs. *Canadian Journal of Public Health, 73,* 259–262.

Siu, A. L., Sonnenberg, F. A., Manning, W. G., Goldberg, G. A., Bloomfield, E. S., Newhouse, J. P., & Brook, R. H. (1986). Inappropriate use of hospitals in a randomized trial of health insurance plans. *New England Journal of Medicine, 315,* 1259–1266.

Smith, G., & Falk, H. (1987). Unintentional injuries. In R. W. Amler & H. B. Dull (Eds.), *Closing the gap: The burden of unnecessary illness* (pp. 143–163). New York: Oxford University Press.

Steiber, S. R. (1987). Right to die: Public balks at deciding for others. *Hospitals, 61*(5), 72.

Stirewalt, C. F., Linn, M. W., Godoy, G., Knopka, F., & Linn, B. S. (1982). Effectiveness of an ambulatory care telephone service in reducing drop-in visits and improving satisfaction with care. *Medical Care, 20,* 739–748.

Tanner, J. L., Cockerham, W. C., & Spaeth, J. L. (1983). Predicting medical utilization. *Medical Care, 21,* 360–369.

U.S. Department of Health and Human Services. (February, 1993). *Cataracts in adults: Management of functional impairment.* Rockville, MD: Public Health Service, Agency for Health Care Policy and Research.

Vickery, D. M., Kalmer, H., Lowry, D., Constantine, M., Wright, E., & Loren, W. (1983). Effect of a self-care education program on medical visits. *Journal of the American Medical Association, 250,* 2952–2956.

Wagner, J. L., Herdman, R. C., & Wadhwa, S. (1991). Cost effectiveness of colorectal cancer screening in the elderly. *Annals of Internal Medicine, 115,* 807–817.

Wennberg, J. E. (1990). Outcomes research, cost containment, and the fear of health care rationing. *New England Journal of Medicine, 323,* 1202–1204.

White, P. F. (1988). Use of patient-controlled analgesia for management of acute pain. *Journal of the American Medical Association, 259,* 243–247.

Wickizer, T. (1992). The effects of utilization review on hospital use and expenditures: A covariance analysis. *Health Services Research*, *27*(4),103–121.

Wilson, J. F. (1981). Behavioral preparation for surgery: Benefit or harm? *Journal of Behavioral Medicine, 4,* 79–102.

Winslow, C. M., Kosecoff, J. B., Chassin, M., Kanouse, D., & Brook, R. H. (1988). The appropriateness of performing coronary artery bypass surgery. *Journal of the American Medical Association, 260,* 505–509.

Winslow, C. M., Solomon, D. H., Chassin, M. R., Kosecoff, J., Merrick, N. J., & Brook, R. H. (1991). The appropriateness of carotid endarterectomy. Santa Monica, CA: *RAND*

Wu, W. C., & Pearlman, R. A. (1988). Consent in medical decision making: The role of communication. *Journal of General Internal Medicine, 3,* 9–14.

Yen, L. T., Edington, D. W., & Witting, P. (1991). Associations between health risk appraisal scores and employee medical claims costs in a manufacturing company. *American Journal of Health Promotion, 6,* 46–54.

4 Rural Health and the New Media

Jane Preston
Telemedical Interactive Consultative Services, Inc.

Health care delivery in isolated areas by interactive video presents both opportunity and challenge. In this chapter two basic problems are identified in the equitable delivery of quality medical services in the United States today. The Texas Telemedicine Project is described as a template of tele-communication technology in cost-feasible service to a rural community, and the dynamics of the most costly problem observed are addressed. Recommendations directed to a national delivery system are then listed.

This is written to encourage careful planning for rural health care delivery. The focus is on building blocks toward economic feasibility as demonstrated in the "living laboratory" of the Texas Telemedicine Project (Preston, 1993).

NETWORK DELIVERY ISSUES

In the United States, increasing numbers of rural hospitals are closing (Texas Hospital Association, 1994). The cost of medical care steadily rises. The gaps are spreading between the quality of care in urban state-of-the-art medical centers and the quality of and access to health care in the remainder of the country. Indeed, the picture is stark in rural communities, but it is equally bleak in urban areas where access is blocked by prisons walls, traffic, enclaves of poverty and ignorance. In short, our country does not provide quality health care equally to all citizens. Delivery of medical services is at crisis level.

Disease and trauma too regularly have the jump on our delivery system. The urgent business of medicine is to gain on these twin foes, and to triumph quickly. Our system does not work well; it is off the mark.

PROBLEMS IN DELIVERY OF HEALTHCARE SERVICES

Demography of Progress

A partial explanation of our delivery crisis lies in the side effects of progress in science and technology. Health care professionals gravitate toward scientific centers of technologies, research, and laboratory resources. This appropriate concentration of interrelated esoteric resources in metropolitan centers, nonetheless, produces a disservice to countless rural communities. Given the spread between communities in most of the United States, distance becomes a hidden ally of disease. A rural community's very functionality and viability are put in jeopardy. Rural families and providers alike suffer from spreading isolation as community hospitals close at alarming rates. So access and quality cluster centripetally whereas rural hospitals, prisons, community health and mental health facilities scatter centrifugally across the land, increasing the gap between experts and those in need of expertise.

Distribution of Disease versus Distribution of Service

Access and quality are further disrupted by a lack of fit between the distributional pattern of medical delivery and that of disease and trauma. The business of medicine is to save life and function when disease and trauma strike. The problem is that the site and timing of disease is unpredictable. Disease manifests itself in manifold forms, frequencies, and locations, and in individuals from varying demographic groups. Additionally, no sooner has an infectious disease been "cornered" by an antibiotic than it springs up elsewhere, too frequently in a mutant form, resistant to previously controlling therapies. Examples are too familiar: AIDS in Africa, malaria in the tropics, trachoma in India, a rising incidence of tuberculosis, diphtheria, and hepatitis in the United States. There are insidious infections everywhere, recurrently slipping from medical control. In addition, noncontagious diseases such as diabetes and melanoma strike individuals with alarmingly unpredictable complications and frequency.

Like disease, trauma and injury appear in individuals at unpredictable sites, swiftly and without warning. They occur through accidents, natural catastrophes, and as byproducts of human ingenuity and the pace of modern life. Automobile crashes claim thousands of lives yearly, mercury toxicity decimates a village, fire sweeps a crowded hotel. Too often, trauma results from inexplicable and unexpected violence directed toward self or others: suicides in young and old, child abuse, gang assaults, and war. Trauma and injury, like disease, confound delivery of care through their unpredictability.

Telemedicine as a Solution

In looking for solutions to the mismatch between appearance of disease and dispersal of expertise, the benefits of a system in which the patient and the local health care provider can instantly consult with an appropriate specialist hundreds of miles away become obvious. To see and hear, to be seen and to be heard, to transmit crystal clear x-ray films, EKG tracings, and patient records instantly, often eliminates the need for a trip by ambulance and the progression of the illness during the trip. Moreover, with suitable training in camera placement and technique emotionality can be preserved, even across distance.

It is not difficult to imagine that a state or federal agency, operating with limited funding, can function more efficiently and serve citizens more effectively when valuable human resources are not wasted in travel time. Unfortunately, wasted travel time is typically the fate of an agency's most precious talent. By using interactive video technology this loss of valuable time, measurable in expertise and dollars, can largely be eliminated, thus extending the agency's useful presence in the community. The highest level of expertise can be promptly delivered where and when needed.

In fact, these public and private benefits are not distant fantasies. Technology, tested service, and cost are now at an affordable level. Since telemedicine's beginning in 1958, decline in equipment and transmission costs have been dramatic. To the mid-80s, costs were prohibitive for a national delivery system, though NASA's *STARPAHC* and 15 other federally funded projects showed that technical quality and access were adequate for many medical services.

In 1980, before research on compression of digital signals became vigorous, one codec (a computer capable of compressing and decompressing digital information) cost approximately a quarter of a million dollars. It occupied two flatbed trucks, and required two days for a five-man crew to mount. (One codec is required at each end of a transmission to translate analog pictures into a digital stream and back to a picture at the other end.) In spring of 1991 one codec cost $45,000; by spring of 1993 one cost about $15,000 to 20,000 with the downward trend continuing apace. Transmission means multiply; copper conditioned, "switched 56" lines, microwave, fiber optic lines, and line-of-sight laser are now all applicable in cost as well as service. All are capable of interconnectivity, even with satellite transmission.

These declining costs precipitated a large increase in the number of telemedicine systems in a short period of time. In 1989, there were only three operating systems: Mayo's, Texas Tech MedNet (now HealthNet), and the Texas Telemedicine Project. At present, approximately 25 states have at least one system which is either active or in the advanced stages of planning (NASA, 1992). The statement of the Council on Ethics and Judicial Decisions of the American Medical Association designated that telemedicine is an ethical means of health care delivery, yet another affirmation of telemedicine's readiness. Clearly telemedicine is financially, as well as technologically, now possible.

Given proper planning, telemedicine's benefits may soon be available to all parts of the country, both urban and rural. However, in order to maximize access, quality and cost efficiency, those who lay the foundation for telemedicine systems must plan carefully to provide economies of scale. For, although large economies are a given with urban massing of people, many small clinics would not require a number of consultations sufficient to attract technological investment.

Fortunately, electronic clustering allows for frugal partnering with other institutions. Through electronic clustering, one small community's health resources can be joined by other rural partners and provide economies of scale justifying investment in the telemedicine tools. The same demographics that produce a lack of medical expertise in rural areas produce a similar paucity in other fields as well. Clustered partners may include industry, education, law, and entertainment providers, all requiring sophisticated expertise in today's world. Interactive video technology could readily deliver management and technical services to industry, law, and entertainment and allow for work and education in the home while still delivering medical expertise. This can revitalize rural areas, bolster economic development, and stabilize and enrich community spirit.

THE TEXAS TELEMEDICINE PROJECT

The Texas Telemedicine Project was designed as the latest in a series of efforts to prepare the American medical profession for telemedical service. Telemedicine, the delivery of services to isolated areas, has been conducted in some form since 1958 (Bashshur, Armstrong, and Youssen, 1975).

I have been privileged to interview professionals whose experiences extended as far back as 1960. Their work, as well as the telemedicine database library, now located at the Medical College of Georgia, has served to inform our own Texas Telemedicine Project.

The Project

The Texas Telemedicine Project is constructed on three concepts: (a) the distinction between the business of medicine (health care) and medical business (financial support structures and management of funds) to enable development of rational decisions about each, (b) acceptance of virtual reality as applicable to a medical services delivery design, and (c) an understanding that economies of scale are required to achieve cost-effective health care in sparsely populated rural areas (see Fig. 4.1, following).

The project had two operational goals. The first was to design a network capable of providing economies of scale to a small community. The second was to test the cost-feasibility of this design. Barriers to and supports of use of the system were, of course, the dynamic forces directing the cost and were foci of the research observations.

Project Design

Delivery system design was evolutional and to be tested through several phases. As of the end of 1993, two phases have been completed: (a) design of a "worst case" cost-feasibility delivery system with community involvement; (b) implementation of the delivery design with concurrent testing and cost analysis; (c) design and preparation of a "best case" scenario analysis, underway; and (d) an extended retrospective study of adoption and dispersal, planned for 1995 (Preston, 1993).

Constraints

In phases I and II several criteria directed choice of rural and urban provider sites:

1. The field sites (i.e., the rural community) should not be robustly prosperous.
2. The rural hospital should be struggling financially.
3. Travel to a tertiary center should be possible rather than clearly impossible thus requiring of users a conscious choice for or against telemedicine use.
4. The tertiary "center" should be heterogeneous in its medical specialties, not an institution, and without "bargains" from a provider group

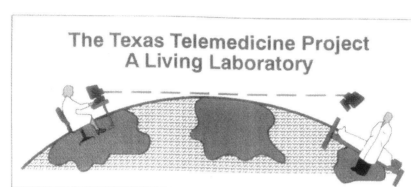

The Texas Telemedicine Project
A Living Laboratory

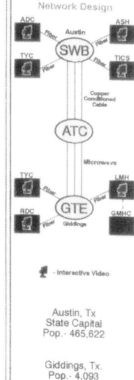

Service

- 2500+ Patient "Visits"
- All Medical Portals of Entry in a Rural Comm. Covered
- Specialties Used Over Network:
  - Allergy
  - Cardiology
  - Dermatology
  - Gynecology
  - Internal Medicine
  - Nephrology
  - Neurology
  - Pediatrics
  - Plastic Surgery
  - Psychiatry
  - Pulmonology
  - Rheumatology
  - Urology

Economy

- 14-22% Net Savings Per Year
- 2.6 Yrs. Payback on Equipment

Network Design

Austin, Tx
State Capital
Pop.- 465,622

Giddings, Tx.
Pop.- 4,093
56 Miles
from Austin

Providers/Users

- Austin Diagnostic Clinic (Nephrology Section), Austin
- Austin State Hospital, Austin
- Texas Youth Commission, Austin
- Regional Dialysis Center, Giddings
- Lee Memorial Hospital, Giddings
- Lee County Mental Health Clinic, Giddings
- Giddings State School, (TYC), Giddings

Supporters

Foundations:

The Meadows Foundation
The Moody Foundation
The Hogg Foundation

Business:

Project Bluebonnet, Inc.
Southwestern Bell Telephone
GTE Central
Advanced Telecommunications Corporation (ATC)
MCC
Video Telecom
Texas Instruments
Telemedicine International
Olympus Inc.
HCA Shoal Creek Hospital
Upjohn Pharmaceuticals

FIG. 4.1.   The Texas Telemedicine Project: A living laboratory.

such as an HMO or PPO (i.e. it is to operate as a virtual specialty center without walls).

5. Equipment, installation, transmission, and maintenance should be at "off the shelf" prices.

6. The transmission should be a mix of regional and long-distance telephone companies producing multiple layers of industry profit and complex interconnectivity.

It is recognized that no one community can be deemed a representative sample of rural America. It is also recognized that aggregate costs, savings, and revenues differ from cash flow management. Immediate cash flow benefits develop as policies are enforced to shift travel savings to system and system investment costs. The immediate financial reaction to the implementation of a telemedical system is increase in fixed assets while lowering monthly expenses. Although many studies will follow, a central problem in definitive economic studies will be inclusion and projections of changes in technology, distribution, and costs of technology, and the culture's adoption and dispersal of change. These are variables not readily computerized into rigorous projects. The Texas Telemedicine Project's notion is simply that if cost-feasible results can be derived with the above design constraints, the findings, as a bottom line study, can be applicable to other communities as well (Preston, 1993).

Field Site: Giddings, Texas

Giddings, Texas is a village of 4,093 people located 65 miles from Austin in central Texas. Giddings was founded in 1871 to provide a service point for a new railroad line running eastward from Austin to Houston on the Gulf Coast. The Giddings station was to accommodate surrounding towns such as Lexington, Fedor, Dime Box, and Evergreen. It is interesting to note that the village was started as a center for transportation of materials; telemedicine now offers the possibility of its being a center for transport of expertise.

Agriculture formed the mainstay industry. In the late 1970s an oil boom brought an influx of people and money to Giddings; cows grazed in pastures dotted with oil field pumps. The boom quickly went bust, however, leaving the town strewn with bankruptcy claims, distress, and a degree of suspiciousness of outsiders. More recently, the town has attracted light manufacturing plants. A handful of lawyers, insurance agents, and bankers have offices alongside small stores and shops. The town has one high school, but no college. Crime is evident, with frequent illegal drug activity, robberies, and assaults. Except for a small opera organization and a village swimming pool, there is no commercial entertainment. During a town birthday celebration, public relations efforts addressed only whether a burial site was indeed that of the last man hung in the village (before the turn of the century).

U.S. Highway 290, on an east-west axis, constitutes the main street of Giddings, Texas, 2 miles in length. South of the main street, near the center of the town, are the city hall and hospital. The 26-bed Lee Memorial Hospital, daily census of 2.6 patients, incurred a debt of $300,000 for each of the three years preceding the project; the community had recently voted in a hospital taxing district to raise public funds. In addition to the hospital, Giddings' portals of healthcare were a renal dialysis center, a community mental health center, and a maximum security youth correctional facility with inmates in need of periodic health care, educational support, parole evaluation, discharge preparation, psychotherapy, elaboration of occupational training, and management support.

Our community analysis began in the first quarter of 1989 and was completed by the third quarter. The contacts included leaders in community politics, health, industry, law, and education. The process included demonstrations and education about telemedicine, interviews for input as to possible applications as well as their interest in participating. Training and troubleshooting were scheduled by interactive video, making Telemedical Interactive Consultative Services, Inc. (TICS, Inc.), available for 30 minutes each day to each component.

Provider Site: Austin, Texas

Austin, Texas has a population of 475,622 and an unusually strong medical community for a city without a medical school. There is a graduate school of nursing at the University of Texas, and the Travis County Medical Society serves as a collegial focus for medical professionals. There are board-approved residencies in family practice, pediatrics, and internal medicine, and rotating residencies in surgery, obstetrics/gynecology, and psychiatry. The Austin medical community structure offers a rich resource in specialty services without producing the distortion in the cost study caused by training stipends, had service been from a medical school, or capitation controls, had service been from an HMO.

Technology Team

Telephone companies serving the project's communities are Southwestern Bell and GTE. Southwestern Bell is centerpoint for Austin's three provider locations and TICS headquarters. These three provider locations and TICS are connected by fiber-optics to Southwestern Bell's Greenway office in

TEXAS TELEMEDICINE PROJECT
SWBT SWITCHING - WEEKLY SCHEDULE
Period: March 11, 1992 UNTIL FURTHER NOTICE   MONDAY THROUGH THURSDAY

TIME	ADC	ASH	GSS	LMH	RDC	TYC	TICS
7:00 am	LMH	off	TYC	ADC	off	GSS	off
8:00 am			TICS			off	GSS
8:45	RDC	LMH	off	ASH	ADC	TICS	TYC
9:00 am			TYC			GSS	off
10:00 am							
11:00 am							
12:00 pm							
1:00 pm							
2:00 pm							off
	LMH	off		ADC	TICS		RDC
3:00 pm		TICS			off		ASH
		off					off
4:00 pm	TICS	LMH		ASH			ADC
	Off	off		TICS			LMH
5:00 pm	LMH			ADC			off
6:00 pm							
7:00 pm	OFF			TICS			LMH
8:00 pm							
9:00 pm							
10:00 pm	LMH			ADC			OFF
11:00 pm							
12:00 am							
1:00 am							
2:00 am							
3:00 am							
4:00 am	RDC	LMH		ASH	ADC		
5:00 am							
6:00 am							

ADC = Austin Diagnostic Clinic
        459-111

ASH = Austin State Hospital
        452-0381

GSS - Giddings State School
        409-542-3686

LMH = Lee Memorial Hospital
        409-542-3141

RDC = Renal Dialysis Center
        409-542-3116

TYC = Texas Youth Commission
        483-5101

TICS = Telemed Interactive Cons. Serv.
        338-3505

GMHC = Giddings Mental Health Center
        409-542-3042

FIG. 4.2.   Automatic optimal switching schedule based upon probable clinical needs.

Austin. Southwestern Bell is responsible for automated switching of a schedule designed for optimal clinical service (see Fig. 4.2, following). GTE is centerpoint for Giddings' user sites, fiber-optic connected. LDDS Metromedia (previously Advanced Telecommunications Corporation) is responsible for long-distance service connecting Southwestern Bell and GTE. LDDS Metromedia connections are Austin to College Station and College Station to Giddings with microwave and copper-conditioned lines, respectively.

The end users' equipment is loaned by VideoTelCom (now VTel), three sets; Jack Moncrief, M.D., three sets (VTel equipment); and Telemedical Interactive Consultative Services, Inc., one set (VTel). The only line installation required for the Project were the fiber-optic lines in Giddings from GTE office to Lee Memorial Hospital, Giddings Renal Dialysis Center, and Giddings State School. End equipment was placed in Giddings at the Renal Dialysis Center, State School, and Lee Memorial Hospital. Staff and patients from the Mental Health Center use Lee Memorial Hospital equipment as the Center is adjacent to the hospital. End equipment is VTel CS350-based with codec, two 20-inch monitors, a pan/tilt/zoom camera controllable across distance, two conferencing microphones, and an electronic graphics tablet; all of which provide fully interactive communications between sites at 1/2 T1 (768 kilobytes) speed. A VCR and tapes are provided at each site. Document stand with camera is provided at Austin State Hospital, Lee Memorial Hospital, and TICS office. One-half T1 transmission is used routinely, though three T1 lines connect all sites around the clock (Preston, 1993).

Provider Plan

The project director chose certified specialists based on availability, rather than negotiate a contractual agreement with an HMO, PPO, or any provider group. This decision was in keeping with the concept of agile medicine, or a clinic without walls, workable nationwide. The Travis County Medical Society roster became the directory of specialty services to the community of Giddings. Telemedicine staff credentials for both field and provider sites were those of the Travis County Medical Society. This credentializing process constituted a test step toward formation of a telemedicine staff, a standard for delivery of telemedical care.

Austin, as the Texas capital, is the headquarters for state government agencies. Among those agencies are the Texas Youth Commission and the Texas Department of Mental Health and Mental Retardation. The Texas Youth Commission, paired with the prison in Giddings, is responsible for management of youthful offenders under 18. Giddings State School is a maximal security facility for those classified as violent offenders. The Texas Department of Mental Health and Mental Retardation, paired with the Giddings Mental Health Clinic, is responsible for the care of the mentally ill and mentally retarded of the state of Texas. The Austin Diagnostic Clinic, a group of 160 specialists, sends nephrologists to Giddings to serve the Giddings Renal Dialysis Center patients. Equipment at the Austin Diagnostic Clinic serves Austin Travis County Medical Society specialists called at request of physicians at Lee Memorial Hospital in Giddings. There were 11 different specialists used in the first year.

Additionally, Austin is home to the University of Texas, including the Lyndon Baines Johnson School of Public Affairs, the graduate school of business, graduate school of nursing, and graduate school of telecommunication, and the microelectronics and computer technology corporation (MCC). Each has provided valuable expertise to the project.

Clinical Considerations

During 2 years of Phase II implementation, more than 2,700 patient consultations took place. The technology is remarkably reliable and user-friendly. Although end equipment training was offered to all, not all users took advantage of the offering. Nevertheless, on occasion, even untrained specialists, called for emergencies, were able to promptly address diagnosis and treatment. Interference with transmission occurred less than 2% of total transmission time. Noteworthy is the fine, smoothly coordinated cooperation between long-distance and regional transmission personnel. These companies had no previous experience working in a medical research environment involving patients. Moreover, neither the equipment nor the transmission lines were designed specifically for this project; those which were available were pulled together quite effectively. Additionally, redundancy was not inbuilt in system connectivity. A human alerting chain interconnecting all four telecommunications corporations with the project director was designed, activated, and currently responds 24 hours a day, seven days a week.

The clinical network schedules are set by the project director. The lines are made available based on probable clinical needs. For example, Lee Memorial Hospital at Giddings is connected to Austin Diagnostic Clinic from 5:00 P.M. to 4:00 A.M. to care for possible victims of a major highway accident on Highway 290, which traverses Giddings. At 4:00 A.M. Austin Diagnostic Clinic's connection is switched to the Giddings Renal Dialysis Center when nurses arrive to prepare for the early shift of dialysis patients. The schedule holds that linkage until 7:30 A.M. to handle dialysis rounds. At that point Austin Diagnostic Clinic and Lee Memorial Hospital at Giddings are reconnected to handle consultations requested by the Giddings physicians. It is agreed that unscheduled switching will occur at the call of the project director, who is responsible for management of emergencies.

DATA COLLECTION

Ultimately, telemedical volume of use determines cost-feasibility, hence a variety of use factors were addressed in data collection. The project addressed the following: user acceptance, completeness of documentation, clinical access and technology reliability, institutional adoption and diffusion of telemedicine.

User Acceptance

One may envision the project as an electronic hospital-clinic without walls. Based upon user acceptance analysis to date, once a site has made a commitment, the providers and patients are almost unanimous in their acceptance of telemedicine. There were three physician and one patient refusal to use the system. One physician who had not had time for training (approximately 10–20 min), was most reluctant to see a patient of his who was requesting to be seen "on the box at the hospital" rather than to drive the 65 miles. He ultimately agreed to see her, if only to persuade her to drive to his office. He quickly began conducting his clinical evaluation, to their mutual satisfaction, using the telemedicine connection. This led to a negotiation for a weekly telemedicine clinic time to see patients from the distant area. Such a shift of attitude in individuals and community defines the acceptance of a clinic without walls.

Two of the physicians (both psychiatrists) refused to learn to use the system because they preferred being in the room with the patient and enjoyed the drive. A third, also a psychiatrist, was a foreign medical graduate who feared loss of her license if she "did anything strange." The only patient's protest centered on a transfer of her usual psychiatrist which she mistakenly thought was because of the use of telemedicine.

Completeness of Documentation

The data sheet (see Fig. 4.3), was submitted at each site for all transmissions. This check sheet documents patient information, satisfaction, travel choices, salary range of those saved travel, equipment satisfaction/problems. At each site there is one person designated for schedules, data management, and so forth. Data sheets are collected and mailed to TICS weekly. Data are reviewed by three people. Though data forms were submitted in 100% of patient encounters, completion was spotty in providing information about

TELEMEDICAL TRANSMISSION REPORT FOR <u>MEDICAL</u> USE

DATE _____

BOTH SITES COMPLETE REPORTS FOR EACH TRANSMISSION

Your Site: _____   Distant Site: _____
Time use began: _____
Time use ended: _____

Purpose: _____
       (i.e. doctor consult, patient evaluation, etc.)

PATIENT WAIVER MUST BE SIGNED BEFORE YOU PROCEED!!!!!!!!!!!!!

VIDEO TAPE ALL TRANSMISSIONS!!!!!!!!!!!!!!!!!

EMERGENCY??   _____ YES   _____ NO

NAMES OF THOSE PRESENT:
     DOCTOR: _____   PATIENT _____
     NURSE: _____   PATIENT CHART # _____
     OTHERS _____
     _____

PRIMARY DIAGNOSIS _____   (DSM code)

SECONDARY DIAGNOSIS _____   (DSM code)

PRIMARY TREATMENT _____   (DSM code)

PHYSICIAN'S NOTES (IF ANY) _____
_____

WITHOUT EQUIPMENT, YOU WOULD HAVE: _____ TRAVELED
   _____ NOT CONSULTED   _____ USED A TELEPHONE

IF TRAVEL WOULD HAVE OCCURRED, _WHO_ WOULD HAVE TRAVELED? ___
_____
_____

PLEASE CHOOSE ONE OF THE CHOICES:

Consultation was: <u>very useful</u> <u>gave equivalent results</u>
                         <u>not useful</u>

Transmission quality: <u>good</u>   <u>fair</u>   <u>poor</u>

After how many attempts: <u>one</u>   <u>two</u>   <u>three</u>   <u>four+</u>

If problems, reported to: <u>individual on site</u> <u>TICS office</u>
                              <u>did not report</u>

Equipment limitations: <u>none</u>   <u>controls ??</u>
       <u>instructions??</u>   <u>picture??</u>
     Limitations explanation, if any: _____
_____

FAX FORMS EVERY FRIDAY TO: 512-338-3600
PROBLEMS?, CONTACT REBECCA AT 512-338-3505

FIG. 4.3. Telemedical transmission medical report.

salary range. In most instances salaries could be derived from role salary ranges. Satisfaction and problems were described to date. There was, and is, limited compliance with the requirement to videotape every transmission. Physicians promptly become involved with a patient, as do nursing and other personnel, and forget to start the taping. As the tapes represent physicians' best protection against liability suits, this is somewhat surprising.

Apart from the data sheets, credentialling of a telemedicine staff presented interesting findings. Only two of four physicians in Giddings completed the credentialling process. The two who did not were older physicians, chronologically and in practice years. Neither cooperated with offers for training, though during Phase I they had expressed support for the project. One closed his practice late in the second year of Phase II.

Clinical Access and Technology Reliability

Access to patient care via telemedicine was excellent and most clinics were on schedule. Rarely did a clinic have to be rescheduled because of equipment or transmission problems. Overall, the electronic system was down less than 2% of the time. Equipment did not provide user control of speed or site selection, nor was there inbuilt redundancy. All equipment dated from 1991, with the exception of three MUX and one monitor, which required replacement.

Institutional Adoption and Diffusion of Telemedicine

The concept of distant delivery of medical services received ready adoption by individuals at the management and user level of each site. Implementation (dispersal), however, was poor in five of six sites. Typically, the decision to use the system came from top management, but was delegated to a subordinate. Sites where implementation was poor conducted no meetings for discussion of goals, opportunities, fit with operational patterns, outcome expectations, measurements, feedback, or other implementation strategies.

Institutional policies regarding telemedicine were established only at the nephrology section of Austin Diagnostic Clinic. In no site were incentives for use and disincentives for driving considered, though expenditures for travel and losses of functional time through such travel were significant. At some sites policies were in place that inadvertently prohibited use of telemedicine. For example, at the Giddings State School, contrary to recommendations, the equipment was put in the administrator's conference room, which was off-limits to the clients. This, of course, also precluded any telemedical service to State School's prisoners ("clients"). The equipment was most often used in terms of emergency management, somewhat less so as a training and management tool, and least as a network communications tool of considerable power to coordinate resources toward accomplishment of an agency's goal. Such inappropriate usage suggests a lack of attention to converting the agency's mission into telemedical policy. There was adoption of the technology as a tool, but no network dispersal.

Only the nephrology section of the Austin diagnostic clinic considered the technology as a tool to increase efficiency in their mission. From the beginning they discussed equipment, schedule, and personnel as related to their charge to dialyze groups of patients at different sites. The discussions were casual, ongoing for three weeks, and included input from all personnel involved. Policies and decisions were promptly reached and tested. They adopted the network concept and then moved to incorporate it into their mission plan.

In assessing this site's salutary ability to use telemedicine so aggressively, several observations may bring light to its successful implementation. First, the nephrologist who most used the equipment had made a financial commitment to do so. Second, the nephrology section regularly works as a scheduled personnel network involving a conscious time-distance focus. Eight nephrologists medically cover four scattered renal dialysis centers, working together well with staff in a smoothly coordinated schedule. Telecommunication tools, such as phones and beepers, are already integrated in the management of their intricate responsibilities and schedule. A third factor in their successful implementation is that they regularly probe for new technology in their specialty work, though none was actively involved in telemedicine until the project.

SUMMARY OF FINDINGS

Findings are summarized in three graphs (see Figs. 4.4, 4.5, & 4.6). The data summarized represent 12 months of service delivery. Network components' cost, savings, and increased revenue are aggregate figures. The increase in combined volume of use each quarter correlates with a growth in economic benefits. The steady improvement is in spite of lack of supportive policies in those to whom major segments of line time were assigned. Nonetheless, given a continuation in the rate of growth, payout would be complete in 2.7 years. Costs covered lines, connectivity, equipment, insurance on equipment, maintenance contract, and installation. Savings were from reduced travel, elimination of redundant tests and personnel, and absorption of audio (telephone and fax) charges. Increased revenue occurred from reinstitution of obstetrics after telemedical retraining of neonatal techniques for nurses and from increased census (i.e., patients who did not require transfer).

Overall, the project components received 95% of the economic benefits.

Graph 1
Overall Profit(Loss) By Quarter

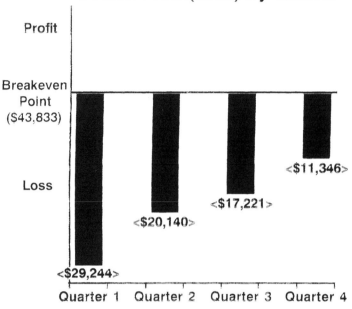

FIG. 4.4.   Overall profit (loss) by quarter.

Overall Use by Quarter

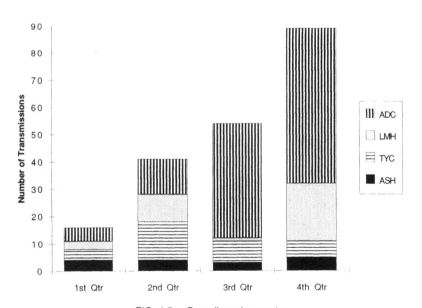

FIG. 4.5.   Overall use by quarter.

One Year of Telemedical Interactive Consultative Services

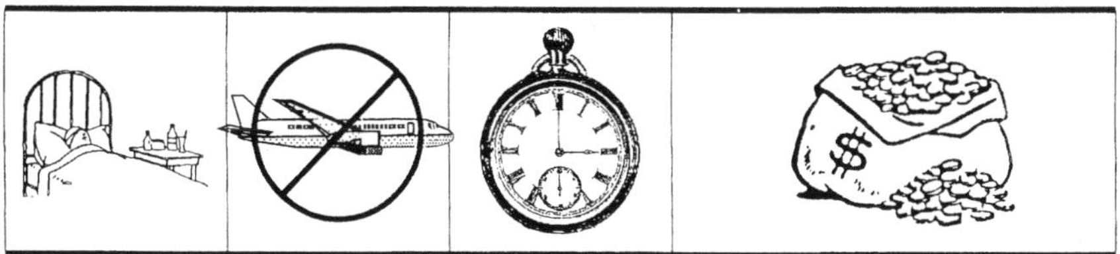

Number of Patients	Trips Avoided	Hours Equipment Used	Cost Feasibility	
			1 year	2.7 year
1,999	302	307	<$-93,317>	break even

FIG. 4.6.   One year of telemedical interactive consultative services.

Of the 95%, 75% was from resource and direct savings, and 20% was increased revenue. Of the 75%, 15% was immediate revenue saving from reduced travel and communication costs; 60% accrued through personnel realignment. Savings from road salary and duplicate personnel are reflected in Fig. 4.6. The patients received the remaining 5% economic benefits through reduced travel costs.

PROJECTS DEMONSTRATING USEFUL CONCEPTS

A number of projects have arisen since 1990. They are impressive in number and variety of approaches. All have important contributions, but three models are uniquely valuable and complementary for national planning.

The underlying tenet of all telemedical projects in the United States is dispersal of health care to those in need. Yet important differences are expressed in the variety of project goals. These differences serve the nation well by presenting complementary templates for rapid, cost-effective improvement of our health care delivery system. All must be viewed against the impressive research of the 1970s which tested concept and equipment reliability, specialty uses, and user satisfaction. In addition, the nearly flawless record of health care of the astronauts in space over the last 30 years is an awesome presentation of distance-delivered medical care. We stand on the shoulders of reliable giants!

The three projects described below do not cover all the innovative work going on presently. However, differences in their goals and the lessons they can teach should be of prime consideration in national planning. They demonstrate both a rapidity in the spread of services possible with strong political leadership and economic and technological sophistication in dispersing education and support. They represent a model of cost accounting, and production of economies of scale, in a national delivery system as well as a viable economic strategy for a profitable telemedicine delivery system. Additionally, they point to a need for research in adoption and dispersal.

Example I.   The Medical College of Georgia is an outstanding example of the value of strong political leadership. Their goal "is to assure that everyone in our state, whether living in the heart of Atlanta or on a south Georgia farm, has immediate access to quality medical care," according to Jay H. Sanders, M.D., FACP, Director, Telemedicine Center, Medical College of Georgia (personal correspondence, June 1994). The political leadership allowed this system to grow rapidly and expand its care and education to 14 sites. Health delivery is strengthened by use of sophisticated peripheral equipment. The decision of the governor of Georgia to work with the president of the medical college was critical. The leadership directed itself toward obtaining funds from an overcharge telephone windfall to the state of Georgia. The funds were assigned to telemedicine for better dispersal of health care to the citizens of the state. The result has been a project with widening influence as one example of a telemedicine health care delivery system.

Example II.   Texas Tech's HealthNet has developed three approaches to education and service. HealthNet's goal is to afford educational and clinical support to isolated health care providers. Texas Tech's four far-flung campuses (Amarillo, Odessa, El Paso, and Lubbock) were first joined as an interactive video base for planning and administering a project. Concomitantly, a service to deliver fetal heart strips, sonograms, x-rays, and pathology slides into tertiary care centers was marketed to small community hospitals. The second approach was contracting with small hospitals to deliver one-way video, two-way audio medical education courses. These courses covered allied disciplines as well as physicians. In the third phase interactive video is being installed in five small communities for educational and clinical support for isolated physicians as well as for didactic courses. HealthNet has obtained national provider organizations' credit certification for a variety of health care disciplines. Continued education now stretches from the Panhandle of Texas to the Rio Grande, 485 miles, and includes some 100 rural contracts.

Example III. The Texas Telemedicine Project, detailed previously, stands as the single template for a system design basic to economic viability in the nation's health care planning. The Texas Telemedicine Project's goal is to design and test a template for cost-effective viable delivery of telemedicine.

These three projects, too briefly summarized to reflect their rich contributions, might also be thought of as different parts of a single, comprehensive, large-scale virtual research project on telemedicine planning. Such a virtual research project would ask such questions as: What are the comparative growth, quality, cost, and access factors present with telemedical services initiated from a tertiary center? What about from a rural virtual reality specialists' center? Can their strengths be complementary and advantageously used? What are patterns, quality, satisfaction, and uses of telemedical education as seen in the following: (a) a tertiary center project with education as prime goal; (b) projects with service as prime goal, though including education; one, an academic tertiary center project, and one, a rural community virtual reality specialists project? Does telemedicine separate centralization of research from training and dispersal of specialists?

All research questions must take into consideration developing sociodemographic changes emerging out of the ongoing developments in the telecommunication technology fields. What sociodemographic changes will interactive video produce in the next 10 years? What of the reverse? Will interactive video use by industry, law, and education deurbanize the U.S.? There do appear to be early trends in that direction. If so, will interactive video maintain a constant number of specialists needed in this country by accelerating the export of our medical expertise abroad? Will there be a differentiation between basic academic resource centers and scattered on-line specialists for consultations? These are challenging questions!

Telemedicine: Computers and Confidentiality

Compression of digital signals is, of course, the most economic means of transmission and dispersal of audio-video in true network fashion. Microwave, laser beam, satellite, copper-conditioned lines, fiber-optic work well, either solo or conjointly. However, it is only through today's codecs, which can quickly and efficiently compress and decompress digital signals, that a truly inexpensive infrastructure can provide nationwide coverage. In the past, microwave, then satellite, transmissions served telemedicine well, early proving clinical reliability and efficacy. Codecs virtually removed economic blocks to service. The delay in dispersal is in national adoption and dispersal of the concept, and in strong national leadership to eliminate expensive political meanderings on the way to a true network.

The computer, in adjunct to audio-video, serves to transport, mesh, program, and order massive data, all necessary to a national network. The research and development work ongoing on a computerized *smart card* should allow any citizen to pay his phone bill, or pull up that portion of his medical record he desires, with one card. Medically, patient's will hopefully have the advantage of such a card that can also pull up visual identification, medical and family history, present and past clinical history, and video tapes of clinically salient information. A monitor to reveal the clinical video should be at patient bedside (home or hospital) to both show past and record present physical disease and function. Such an arrangement allows for ongoing, fully knowledgeable visual consultation between patient and health care provider, for direct instruction to laboratory and x-ray, and for immediate visualization of unusual findings in the laboratory. The information could thus be immediately available, instructions and findings directly given in a timely manner. Transcription mistakes are a constant menace in our medical system. The combination of audiovisual and computer used in immediate fashion with patient visualization and hearing of the process should significantly reduce such errors.

As a national ethic and goal, present efforts to assure confidentiality are laudatory. As a physician, and occasional patient, I applaud the effort. However, as a physician, I must observe that the public is, in action if not attitude, relatively indifferent to confidentiality issues. Insurance forms are quickly signed that allow total chart inspection by numerous clerical workers. Laws state that patients have ownership of their medical information, to be held in the protective hands of the physician; however, laws also perforate that double patient-physician shield so that multiple legal reasons supply subpoena power to pull a full chart out of the physician's files. Many years' experience in major extraction industries' use of secret and sensitive computer information has proven that there is a fairly high degree of computer-transmission confidentiality possible in business. Their techniques can surely be applied to national information about individual citizens. If, of course, that is the will of the people.

RECOMMENDATIONS

Basic research on adoption and dispersal should be strongly supported. There should be care taken to include ethnic and demographic variables of the United States and potential foreign markets for transmitted expertise.

Accounting methodology to document travel expenditures and cost of expertise lost in travel should be designed, standardized, and implemented in all public institutions, not just health institutions. Where identifiable, potential cost of lost opportunity should be reported.

Barriers that prevent interactive video access to homes as well as public institutions should be addressed, be they industry-based, provider-based, consumer-based, or based on state or federal regulatory strictures and lag. Reimbursement for telemedical care should be promptly implemented.

Marketing studies regarding public receptivity to various educational approaches should be done in a heterogeneous cross section of the various environs of the United States (Indian health reservations, retirement and nursing homes, homeless shelters, inner city ghettos, prisons, for example).

Joint telemedicine projects by department of environment, department of transportation, environmental protection agency, department of defense, housing and urban development, and department of health and human services should be concomitantly designed to evaluate telemedical impact on energy conservation, air quality measures, redeployment of government workers, reduction in travel, and cost-effective delivery of services.

The American Telemedicine Association should pursue leadership in producing an expert panel of physicians and allied disciplines to evaluate clinical impact of new telecommunication technologies annually, including any need for updating clinical practice guidelines. Also recommended would be a medical-technical board to investigate evolving teletechnologies with potential applications in medicine, maintaining a dialogue and working in tandem with the panel described previously.

Nationwide provision and standards for redundancy, user controlled dial-up (multisite, multispeed control) should be instituted.

The video evidence of all health care video encounters should be owned by the patient, archived so as to be fully recoverable, and include videotaped salient clinical diagnostic and treatment segments. Patient/citizens should designate, unless mentally incompetent, what is to be "locked" confidential material, to be released only with written or videotaped patient/citizen permission. Locked information shall be reviewed for renewal every three years along with renewal of driver's license and video identification. It shall be the responsibility of the health care provider to make clear to the patient any medical hazard that could accrue from locking specific medical information. This warning should be a part of the videotaped record along with a videotaped acknowledgment that he/she understands. The decisions and responsibility for what is to be locked should rest with the patient/citizen.

Software for clinical records should have the capability to present audio-video, graphics, and text, and the telemedicine format should be capable of 2-way video, graphics, text on screen simultaneously, and of freezing an image for reciprocal distance notations during consultations.

Telemedicine staff credentials should be instituted. A credentials committee should serve according to usual joint commission on American health organizations procedures, including publications of requirement, recredentialling and due process. Continued medical education requirements should be designed for practitioners using telemedicine.

Telemedicine should be fully exploited for preventive medicine and the training of individuals toward full comprehension of the vital part each has as a full member of any diagnostic and treatment team. Disease and trauma battle within an individual; it must not be forgotten that this individual is the sole expert on how the disease, or trauma, is being manifested and on treatment effects. Preventive medical programs should include patient education in observing and describing symptoms.

Sociopsychiatric programs at all levels of education should be presented to prepare the citizenry for ongoing change. Early specific focus should be to support, guide, and negotiate shifts from departmentalization to "orchestra" organizational patterns. All participants should be trained and rewarded for creative think-tank approaches to mission-oriented telecommunication network design and use. Example: one bank offered "frequent nonflier" awards for use of teleconferencing rather than travel in training and management. Training should utilize techniques to support transitions in perceptions. Example: link change and projections to somewhat similar successful past experiences, such as how it was five years before and after the dawning of the present fax age. Leadership should be provided to carefully guide through the transition, and should be made available through tapes and manuals, as well as through personal and/or audio/video contact.

In the interest of health care and monetary savings, simplistic projects and investments to prove telemedicine works should receive no further public funding, considering 36 years of sterling performance.

Reimbursement for telemedical provider services utilizing standard CPT codes and an alphabetic suffix should begin immediately, allowing further comparative research with national samplings.

REFERENCES

Bashshur, A., & Youssen. (1975). *Telemedicine: Explorations in the use of telemedicine in health care.* Springfield, IL: C. C. Thomas.

National Aeronautics and Space Administration. (1992). *STARPAHC* [Videotape]. Clearlake, TX.

Preston, J. (1993). *The Texas telemedicine project: A living laboratory, phase II.* Meadows Foundation Report. Austin, TX: Telemedical Interactive Consultative Services, Inc.

Health Care as Teamwork: The Internet Collaboratory

G. Anthony Gorry
Rice University

Linda M. Harris
Center for Health Policy Research, George Washington University

John Silva
Advanced Research Projects Agency

Joseph Eaglin
MACRO International

> *... the probability that exactly the right collection of expertise will be collocated is small.*
>
> —Wm A. Wulf, 1993, p. 261

The specialization that has been one of the hallmarks of the American health care system has brought the country medical expertise of the highest order. But from another vantage point, this specialization translates to uncoordinated health care services for Americans.

Consumers want integrated services, and a very few institutions and agencies actually coordinate services across organizational boundaries. But on the whole it is difficult to knit providers and patients—separated as they are by geography, organizations, lack of knowledge, or even, ironically, by the boundaries of their expertise. For the poor, the sick, the jobless, and the homeless, these separations create inefficiencies, duplications and gaps in accountability, and parochial and fragmented systems of health care, community development and housing, family services, and job training.

In this chapter, we consider the ways in which advances in computing and networking can enhance the integration of health and human services by interconnecting health experts so that they can collaborate across time, space, and organizational boundaries. New technologies for collaborative work have the power to reshape not only the delivery of health services but the very structure of health services.

Although advancing information technology enables significant organizational change and development, the integration of social and health services depends on a complex interplay of strategies, people, structures and technologies across organizations.

Viewed narrowly, an organization's strategy deals with the way in which it defines its business: What does it do and for whom? More broadly, strategy encompasses the ways in which an organization shifts its direction, perhaps remaking itself in a significant way (Mintzberg, 1989). Efforts to integrate services across organizational boundaries invoke strategic questions for a hospital, an agency or a job-training center.

But strategic intention alone does not determine organizational development. People's skills and aspirations, their work patterns, and the tools they have determine the extent to which their strategic aspirations can be realized. The management challenge is to balance an organization's strategy with the capabilities of its people, its structure, and its technology. This challenge is compounded when a strategy must encompass more than one organization, as generally would be the case in the integration of health and social services.

The authors of this chapter are engaged in the development of networks that foster professional collaboration in a variety of settings. Here we emphasize the role of technology for collaborative work in health and social services integration. We consider the use of networked multimedia technology at the point-of-service (the interaction between the provider and the patient), because this is the defining activity of services provision. We consider how the new technology permits reinvention of the process by which services are selected and delivered.

But we recognize that a complex interaction of factors affects the extent to which an organization can respond to the profound changes in the nature of work induced by computing and its related technologies. Organizations that are most likely to succeed are those that link their information technology strategy to the essential purposes of the institution. This linkage, in turn, depends on leadership and vision to motivate and sustain the entire effort.

In this regard, we enumerate some principles for building and sustaining network collaboration. Consideration of these principles goes beyond the technology to matters of human communication that assume particular importance in these new, distributed collaborative environments.

As an illustration of the potential power of this new technology, we discuss the work of a partnership in Washington, DC that includes technology companies, a social services consulting firm, two universities, a coalition of social services agencies, and governmental entities with broad responsibility for systems of health care and social services. The principal aim of the partnership is to exploit distributed multimedia computing to reinvent the way in which human services are delivered.

THE ADVANCE OF TECHNOLOGY

Advances in computing technology will be the key to a superior coordination and collaboration among providers, agencies, caregivers, and families to integrate medical, health, mental health, social, and economic support services. Computing and information systems will be constructed from increasingly powerful, mass-produced, inexpensive, small components: semiconductors, electronic packaging, circuit boards, disk drives, lasers, scanners, and displays. One consequence will be the "digitization of all information industries and their convergence into a single information hardware sector based on a common set of digital component technologies." (Ferguson, 1990, p. 58).

Through this convergence, workstations will bring together a number of digital technologies, including digital video. For the health care provider, the same workstation can serve as a scanner, fax machine, video-conferencing port and telephone. The user's interaction will be through keyboard, touch, pointer devices, and speech.

A wide variety of knowledge will flow to these new workstations over high-speed networks. Many of the novel applications of information management and presentation we envision today presupposes an ability to intermingle video, movies, sound, print, and display; an ability that is founded on a shared, digital foundation. High-bandwidth networks, a prerequisite for the realization of this workstation vision, will be widely available within the next 10 years.

Already the national networking infrastructure is highly developed and can serve as a foundation for a new kind of collaboration in the provision of health and social services.

The benefits of this infrastructure are already apparent in a wide range of scientific endeavors. Different scientific applications call forth different configurations of the underlying technologies to give users a dramatically more responsive information environment.

THE INTERNET

The centerpiece of the national networking infrastructure is the Internet, the descendant of the ARPAnet, an experimental network designed some 20 years ago by the Advanced Research Projects Agency (ARPA) in the department of defense. Its original purpose was to support military research—in particular, research about how to build networks that could withstand partial outages (like bomb attacks) and still function. Much of this research was conducted at universities, where ARPAnet spread from military research departments to other disciplines.

During the late 1980s the National Science Foundation (NSF) provided significant support for the ARPAnet by creating five supercomputer centers. These centers were the first to provide fast and powerful computer support to researchers outside the military and corporate communities. Using ARPAnet's technology and standards, the NSF built networks to connect these centers. In time, the name of the network was changed to Internet, reflecting the growing usage outside the military research community (National Research Council, 1994).

Now, not only are most campuses connected, but government agencies, nonprofit organizations, and businesses can have access to each other. Even individuals can have limited access to the Internet from home through commercial gateways. There are an estimated 15–20 million Internet users today.

Through the Internet, members of distributed organizations have markedly changed the ways in which they manage their collaborations within their own organizations and across the country. Individuals and groups gain much better access to information on the programs and accomplishments of other parts of the organization.

We briefly review the role of networking in health and human services research in the following section.

NETWORKING IN BASIC SCIENCE

The successes of science over the last decade have dramatically changed the process of research and development. Where an individual might once have spent a lifetime working in isolation on a particular scientific problem, now whole institutions work to solve the same large problems. Although competition is intense, research results are, to a very large extent, freely exchanged (Brunet, Morrissey and Gorry, 1991).

In many scientific disciplines, research has become a group effort. Groups may be located across continents or within a single lab, but the essence of modern science is the interaction of groups and shared information.

James Watson, whose collaborative work with Francis Crick on the double helix won them a Nobel Prize, attributes the collaborative aspect of his work to much of his success, "Nothing new that is really interesting comes without collaboration" (from Schrage, 1990, p. 34). In fact, in a review of approximately 41 Nobel laureates, sociologist Harriet Zuckerman concluded, "Nobel laureates in science publish more and are more apt to collaborate than a matched sample of scientists" (from Schrage, 1990, p. 46).

The development of networking technology, particularly wide area networks and the Internet, has given scientists new tools for collaborative work, and in many cases has enriched their intellectual environments. Distributed organizations, linked through networking, have emerged to confront some of the most compelling challenges of science and engineering. Similar innovations can transform our health care system.

Expansions in the capacity of some sections of the Internet to carry and store multiple media are making it easier to conduct collaborative basic research among scientists, distributed throughout the world. They are building virtual laboratories and collaboratories, that are becoming more and more akin to working around the same lab table (National Research Council, 1993). At the virtual lab, researchers who share the same research question, but who happen to be separated geographically, collaborate in scientific work.

These early collaboratories are typically made up of members of a single basic research specialty who share the same scientific languages and customs, and are in the familiar territory of computers and distributed work environments. Some of these basic scientists have been collaborating over the "net" since its initial days as ARPAnet. With support from the National Science Foundation, ARPA, and others, they have become comfortable using the Internet to form virtual basic research laboratories.

The Human Genome initiative is a paramount example of such large-scale, collaborative science. Teams of scientists, in this case from around the world, are working together on a complex task at the forefront of new thinking.

Because distributed research environments have distinctive characteristics, some of the initial research conducted within them has involved the building and testing of distributed research methods as well as some tools and protocols for facilitating their own distributed collaborative research (National Research Council, 1993).

NETWORKING IN APPLIED RESEARCH

Internet collaboratories for applied research are in their infancy. Where the early, basic science collaboratories have connected members of the same homogeneous research specialty, applied research collaboratories are forming across lines of specialties, bringing their participants a variety of cross-cultural experiences.

In health care there are many areas in which applied work might be conducted distributively. We are particularly interested in the opportunities for health care providers, once isolated from each other by specialty and geography, to collectively explore how to reform the current ways of providing care.

In collaboratories, established specifically for this purpose, providers are joined by computer and networking engineers in much the same way as homeowners now can work with the architect and builder of their home. In the case of health care providers, the language and customs of engineers are foreign and the spaces, processes, and tools they are helping to design are strangely dynamic.

These interdisciplinary collaboratories do not attract the faint of heart. Rather, participating software, hardware, and networking engineers must work diligently to understand how health care providers see their endeavors, share their engineering work space with those who are not professional engineers, and put their designs out for vigorous review and criticism.

Participating health care providers must commit to reform the way they work, to share their work space with strangers, and to learn to work on the network. This may prove as difficult as taking play polo lessons while learning to ride a horse.

The challenges of working in distributed organizations are to a large degree the same as those of collaborative work in single organizations. Teams must share information and task responsibilities in an effective manner. When teams are dispersed across institutions, however, the mechanisms members use to share information and coordinate activities need to overcome the impediments of time and distance. The integration of the organizational components of these centers with the growing national network, the Internet, greatly facilitates this integration, through for example, the provision of electronic mail and file-sharing services.

Although these early health collaboratories are offering us a wild ride, we believe that they provide a glimpse into the future of a seamless system of health care; when in which health care providers—regardless of specialty or locality—can work as teams, each configured for an individual patient's needs.

How do health collaboratories begin? Imagine four health care providers and four engineers from around the country, each sitting at his or her own desk with a personal computer and an Internet connection.

We might see a mental health provider, a trauma care specialist, a pediatric nurse, an orthopedic surgeon, and a software designer, a software engineer, a human-computer interface developer, and a network developer.

They all agree that providers could work more effectively if they could collaborate—the pediatric nurse at Howard University Hospital could help the mental health provider at a church-sponsored program in the inner city of Washington, DC offering counseling to families with children in the last stages of AIDS; the trauma care specialist in the battlefield could benefit from immediate consultation with the orthopedic surgeon at the Army Medical Hospital at Walter Reed.

The two sets of groups—the tool builders and the health providers—begin to discuss how to get from *here*—where health care specialists are isolated from each other to *there*—where health care providers are teammates.

ENVISIONING THE FUTURE

When teams come together in this new networked environment, they will encounter a new context for their work. As DeGregori notes, "The context for the new technology will not be the world as it is today, but the world as it is becoming" (DeGregori, 1985, p. 38).

For health care teams and organizations to gain the greatest benefits from collaborative technologies they must often rethink the ways in which they work. Their response to advanced computing must be dynamic and opportunistic to adapt organizational structures and processes to new technology.

But it is often difficult for health care providers to appreciate this interaction between technology and work—to come to a new concept of collaboration—early in the systems design process. At Rice University, a participative planning process has been developed that emphasizes the interaction between technology and work. The aim of the process is to create a shared vision of the "system after next."

This idea is analogous to the "solution-after-next" discussed by Wadler and Hibino. (Wadler, 1990, p. 135)

The System-After-Next

The system-after-next (see Fig. 5.1) involves a balanced orchestration of technology, people, and processes. It goes beyond limitations and constraints of current systems to recreate an aspect of the organization to take full advantage of current and emerging technology.

The system-after-next exploits technology that is in hand or that will be available in the immediate future, ensuring that technological inadequacies alone are not obstacles to its realization. It serves as the key to a user-centered development process in which the planning starts from the future and moves back to the present.

The schematic above depicts the process through which we link information technology strategy to the target business functions. The process, called iterative prototyping, appears simple, but is in practice a highly interconnected set of activities. In iterative prototyping we take full advantage of

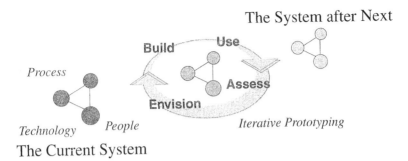

FIG. 5.1.   The community services workstation prototype.

advanced information technology to create increasingly refined realizations of our vision of the system after next. Through successive cycles of prototype creation, process analysis and revision, and organizational assessment, we extend and enrich our concepts, design and implementation.

Implementation is a central element of our activity. We build increasingly better approximations of the system after next. Each cycle of iterative prototyping yields a working version of the desired system that participants in the process can exercise. We rapidly create a framework within which the client can contribute increasingly broad and deep knowledge about the functions and processes under study.

To help groups envision the system-after-next, designers create a video, typically from 7—15 minutes long, in which actors use the prototype to demonstrate the new way of working. These videos, showing people working in new ways, are powerful stimuli to creative thinking by users about the nature of their work and their organizations.

At every cycle of iterative prototyping, management reviews a working system. Because our technology architecture is flexible and open, the increasingly refined prototype points to a smooth evolution of the technology support for the process in question. And we have found that because the prototype is available for use and criticism, it stimulates new cycles of strategic thinking about business operations.

Iterative prototyping culminates in a version of the desired system that can be demonstrated to a range of potential users. It stimulates expansive thinking with respect to the role of information technology in their futures, and in marshalling management support for innovation within their organizations.

Through this process, software engineers at Rice University have created videos that envision the system after next in a variety of collaborative settings: environmental remediation, K-12 education, the coordination of health and human services (see the following case study), the use of electronic libraries, and a number of business problems.

THE TECHNOLOGIC ENVIRONMENT

To facilitate the definition and implementation of new skills and behavior, we need a powerful information technology architecture for the participative design of community health and social services. At Rice, we are developing the electronic studio to help us support a range of interactions between people and technology and to mediate interactions between people.

Our aim is to make the electronic studio a multifaceted environment in which users can rapidly build prototypes and scenarios of the future.

The electronic studio is intended to enhance interactions between teams and the systems that support them. It is expected to facilitate the participative, multidisciplinary design of a health care system—the system-after-next. It will use the same architecture to build prototypes of a new, highly responsive, computer-aided health care system that supports distributed collaboration among providers and between providers and their patients.

Like a movie, recording, or architecture studio, the electronic studio is an environment for collaborative work. But unlike other studies, the electronic studio is not bound to a single place. It can encompass users and multimedia resources across the Internet (or any network).

The stage, sets, models, and props of the electronic studio are built from complex webs of text, graphics, video, and audio. These resources might be located at various nodes throughout the Internet. Its productions retrieve the resources, organize them in novel ways, under a variety of display metaphors, to serve the needs of multidisciplinary teams made up of providers and system design engineers.

Within the electronic studio, it is possible to transform written theories into demonstrations; and models, into simulations. Audio annotations can be heard; and musical scores, played. Digitized video for demonstrations and conferences can be an integral part of collaborative work.

Advanced programs serving as agents could perform functions similar to those of support staff who manage detailed information that is subsequently recast to support higher levels of decision-making.

An important element of the electronic studio is the virtual notebook system (VNS); technology for collaborative work that we have developed over the past five years.

As its name suggests, the VNS is intended to perform many of the functions of an ordinary paper notebook. Indeed, our original idea for the

VNS was stimulated by the prevalence and usefulness of paper laboratory notebooks in biomedical research groups. Using advanced technology, we transcended many of the limitations of pencil and paper, enabling members of a team to share and integrate information in their collaborative efforts. (Gorry, Long, Burger, Jung, & Meyer, 1991)

The VNS is a distributed multimedia hypertext system through which users share notebook pages displayed on workstations. Users can place different kinds of information on these pages including text, images, audio, and even video.

Users can also tie pages together in a variety of ways through the place-ment of pointers called hypertext links. Through hypertext, for example, a team member (a scientist, teacher, or health care professional) can create several paths from a given page for other team members to follow. By invoking a particular link, a colleague is immediately brought to another section of the notebook.

Other hypertext links give immediate access to programs and computers outside the VNS. For example, a group member might create such a link to a library catalog, a research database, a hospital information system, or a collection of artistic images for use in the work of a team.

As long as we deal with homogeneous groups, common tools and organi-zational paradigms such as the VNS are sufficient. However, new forms of collaborations may demand a more diverse set of models or metaphors for manipulating information.

For a potentially wide spectrum of providers in diverse health care settings, the metaphors for information organization and display may prove crucial to the success of the design process. For example, in a community health clinic setting with hundreds or thousands of users, the information system must be essentially self-documenting. It must help its users, many of whom will not be technically sophisticated, understand the interconnection of information resources and service protocols. The same information must be recast in a variety of ways to make it readily understandable to different users.

To facilitate collaboration among heterogeneous user communities, we are extending the electronic studio to accommodate differences in the way collaborators organize and view information, and in the ways they want to interact with information systems.

Many of the features of the electronic studio that are needed to support design are essential for collaborative decision support systems; for example, the management of distributed multimedia, information filtering, and flex-ible information display.

This idea of using the same environment for both the design and the implementation of a system has been put forward before, notably by Englebart and English (1968). But advances in technology and our experience with the design process described above put us in a good position to create significant innovations.

THE COMMUNITY SERVICES WORKSTATION

We now consider the ways in which networking can enhance the integration of community services—even across organizational boundaries. Through a case example, we show that technology for collaborative work, mediated through the concept of the system-after-next, has the power to reshape health service organizations in the coming years.

Complex, contemporary problems such as infant mortality, teen pregnancy, perinatal drug abuse, pediatric AIDS, the treatment of juvenile diabetes, and fetal alcohol syndrome cannot be adequately addressed without a tightly orchestrated provision of services. The medical, health, mental health, social, and economic support demands of these problems require a superior coordination and collaboration among providers, agencies, caregivers, and families, which must be consistent and reliable. For these problems, managed care is not an option, but a necessity. One-stop shop is not a place, but a system of care that envelops all available resources of service delivery and assistance.

Although there is widespread endorsement of such proposed efforts as managed care and one-stop shop service delivery, the more difficult task in most communities is to build an infrastructure that supports such coordination with a holistic approach to service and care.

Our illustrative case demonstrates the way in which networking can be the foundation for the needed infrastructure. For our example, we draw on the work of a partnership in Washington, DC that includes technology companies, a social services consulting firm, two universities, a coalition of social services agencies, and governmental entities with broad responsibility for systems of health care and social services. The principal aim of the partnership is to exploit networked multimedia computing to reinvent the way in which social services are delivered.

In 1991, with a very small grant from the Kaiser Family Foundation, a handful of public and private health and human service providers in the Washington, DC area formed the Health and Human Services Coalition of the District of Columbia.

Each of these providers had been struggling in relative isolation against complex, contemporary health problems that could not be adequately addressed without a tightly orchestrated provision of services.

The coalition sought to build a human infrastructure that facilitated the coordination of services across agencies and institutional barriers. The development of this infrastructure involved the organization of the diverse

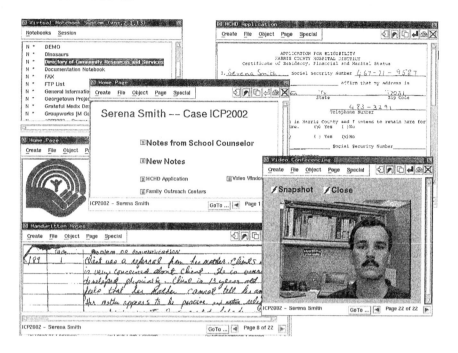

FIG. 5.2.   The CSW Prototype

agencies and providers in the community into multidisciplinary teams. It required guidelines and protocols that supported collaborative case management and team problem solving.

During the formative period of the coalition, they were introduced to a team of software and information engineers who had been developing the electronic studio environment at Rice University. The introduction took the form of a video, produced by the engineers in concert with service providers in their own community of Houston. The video embodied a vision of a system-after-next for the coalition in Washington, DC.

The video envisioned the community services workstation (CSW), connected to other such workstations via a community-wide network (see Fig. 5.2). It would give the provider a variety of tools (including video teleconferencing) to access information from around the community with which the provider and the patient could develop a plan of care. Using the collaborative tools of the CSW, the provider and the client could accomplish in one visit what currently takes weeks.

When the video was shown to the coalition, they invited the team of engineers to join their planning process and technology with the coalition's work on human infrastructure development. This association produced a new information technology architecture that is tightly coupled with organizational development as suggested by the schematic in Fig. 5.1.

The team-building and collaborative case management directed the development of the interagency protocols and supportive software. The protocols and software, in turn, directed the design and organization of the communications technology. Collectively, these processes emerged as a framework that enabled robust transmission of creative ideas and solutions.

Rice University, Howard University, and Macro International have been designing this new information technology architecture for the delivery of community services. With the assistance of Bell Atlantic, the team has assembled elements of this architecture in a demonstration laboratory at the Howard School of Social Work.

The laboratory at Howard, which has been supported by the office of the assistant secretary of health in the U.S. public health service and the National Library of Medicine, contains prototypes of a wide-area network called the Community Services Network (CSN) and the CSW for information acquisition, sharing, and management in the community.

The CSW prototype supports the dialogue between provider and client concerning the latter's needs and the available resources to meet them. Clients must be able to describe their needs or problems as they have experienced them. They also need access to a wide array of information on services for decision-making. Providers, as well, need to compile, sort, search, retrieve, and analyze data in an expeditious manner to assure that services can be obtained as quickly as possible. The multimedia components of the workstation address both the concerns for greater client participation and provider efficiency.

The CSN enables members of a case management team to coordinate across time and distance to address a client's problems. Team members can share assessments of the problem and review collaborative actions. The availability of case management or agency support resources may be posted on the bulletin board as well as alerts regarding funding requests, deadlines, or changes in regulations, or program guidelines that may impact on service delivery.

Such capabilities are common in networked environments such as the Internet. Other components of the CSW are more specific to service coordination and case management.

Coordinating Information Access

Here we review some of these more specialized functions that have emerged from our collaborative design process. Networking enables the realization of each.

One of the major problems in coordinating services across agencies and across care delivery modalities has been the lack of commonly accepted data on every client or family seeking services or assistance. Each agency compiles its information regarding the client's needs, social, health, and financial background, and previous service utilization. There is considerable overlap and duplication in the intake process, in the forms used by various agencies to initiate the service delivery process.

The CSW can free service delivery networks from the rigidity of these forms. We are developing a universal intake function for the CSW that could be used by any agency or provider in the CSN.

The client will benefit greatly from the sharing of information among agencies over the CSN. For example, a case manager can enter most of the information for Medicaid, SSI, and WIC at the CSW; using the CSN to communicate with case reviewers at each of these agencies, while the client is still at the clinic, to refine ambiguous entries. In most cases, this will eliminate the need for the client to endure three repetitive and often tedious intake procedures. Further, the inconsistent information among agencies that often causes delay, rejection, or under prescription of services or entitlements can be greatly reduced.

Sharing Expertise

There are more than 35 entitlement programs in the District of Columbia to address the needs of families and individuals seeking health and social services assistance. Most providers have in-depth knowledge of one or two, but it is virtually impossible for a single agency or provider to be proficient with and keep current on the complexities and continuous changes in the guidelines and regulations of all these programs. Consequently, clients are critically underserved and entitlement resources are often not provided in a timely and appropriate manner.

For the past two years, the health and human services coalition has been working with the united seniors health cooperative, which has developed a decision support tool called the Benefits Outreach Screening Service (BOSS). The program is designed to screen clients for their likely eligibility to receive benefits from most of the entitlement and service support programs, and must be tailored to the guidelines of the city or county in which it is applied.

Access to BOSS across the network enables providers to ensure that clients obtain the services to which they are entitled. Changes in eligibility requirements, reflected in BOSS, can be made available immediately to all providers on the CSN.

Shared Documentation of Cases

Another approach to capturing information or documenting client backgrounds, needs, and experiences is to engage the client in the drawing of pictures or diagrams of their social and familial relationships and the types of needs, problems, and service utilization experiences associated with these relationships. It is called eco-mapping because it allows the clients to describe the environment and the network of relationships and personal experiences that constitute their living conditions.

The CSW prototype demonstrates how a caseworker can draw pictures of relationships and annotate these pictures with text, describing amounts of services needed or received by the client or persons in the client's family, the agencies used, or the problems encountered in acquiring services (see Fig. 5.2).

One of the more difficult tasks in completing assessments of a client's health and social services needs is defining the services to address those needs. Before an appropriate provider can be selected, the precise type, amount, or level of service must be determined to assure that the client's needs are adequately met.

Through networking, we can provide current information on services and providers as well as guideline on eligibility and use.

For each of the selected services, information repositories on the network can indicate how particular services should be used, that is, for persons who have completed inpatient therapy, or for children with multiple disabilities. Such guides can be continually updated to reflect changes in service conditions or the language and definitions used to describe a service or need.

Agencies can broadcast changes in their status over the CSN, and those changes can be immediately entered on the provider profile. Caseworkers may also schedule appointments using the information on the profile.

As clients follow a service plan, each agency can use the network to record their actions, allowing the initiating caseworker to follow the client's progress. This eliminates the need for case tracking, as the networked case file contains all the pertinent information and actions relating to the client. Unlike paper documents, this is a live record of the status of the client and the activities of the agencies that serve the client.

Information sharing over the CSN can help caseworkers address important evaluative questions. Did the client(s) consistently follow through on the prescribed service plan? Were the client needs assessments proven accurate and were the prescribed services appropriate to the needs? Were the agencies selected to carry out the service plan appropriate to the needs and special interests of the client? Through individual and collaborative efforts, case workers and clients may review client files that contain the comprehensive information on all the service activities relating to a client, a family, or a group of clients.

Video Over the Network

As network capacities increase, video sources can play an increasingly important role in the CSW. From network repositories, caseworkers can call up video vignettes depicting situations similar to the client's. This mirroring technique can help clients, particularly with teenagers, to communicate the complexity of their problems or to share information on embarrassing or painful experiences. Video can also give visual representation to types of services, treatments, or care modalities, allowing a more informed selection of a provider to address a particular need.

The CSW prototype uses a number of such videos. The Video Action Group of Washington produced a series of interviews with inner-city teens regarding relationships, sexual behavior, pregnancy, and parenting. Another set of videos, produced by the DC Women's Council on AIDS, consists of interviews with young women who are HIV positive. The women describe their experiences with the disease and how a program called "Sister Care" has helped them to develop a constructive approach to managing their lives.

In another application of networked video services, teleconferencing allows the provider and the client to consult directly with another agency or expert regarding the client's needs. They can make critical decisions that demand a uniform understanding of the nature of the problem and how it should be addressed. Visual contact allows the provider in the initial agency to introduce the client to the other care providers or consulting agencies and establish relationships. This helps clients to more proactively participate in the process of resolving their problems.

Teleconferencing also supports the case review process. The work of interdisciplinary teams conducted by the HHSC required that the teams meet weekly for hours to review cases. By combining distributed computing with teleconferencing, a team can come together at any time, as an assessment emergency arises, without leaving their offices. The providers can think and act as one agency, with a focus on the needs of the client and not the barriers and limitations between their respective agencies. Video telecommunications can become the key to realizing the goals of one-stop shopping and services integration initiatives.

Assessing the Impact

The CSW and CSN prototypes enable us to explore these new, collaborative configurations of technology, people, and processes. With the CSW and the CSN, teams of providers can transcend the obstacles of geographical dispersion and organizational separation to create coherent plans and processes of care for their clients.

The response to our prototypes by representatives from various cities, the federal government, and health care providers has been overwhelmingly positive. This enthusiasm for the CSW and CSN reflects the widespread recognition that health care professionals cannot adequately address complex social problems with disconnected organizations and fragmented services.

As our work proceeds, we look for its impact along both technical and organizational dimensions. We can assess the technology in traditional ways; emphasizing performance, features, and openness of the components of the system.

The more complex and interesting questions arise in the sociocultural domain—in the interaction of the system with people and their work. Here evaluative work might include characterizations of changes in the social organization induced by the technology.

For example, Zuboff recently described the shift toward abstraction of work that accompanied the widespread deployment of computing in several manufacturing settings. In the affected plants, computer mediation of work qualitatively altered the relationship of workers to their environments and undermined conventional work processes. Workers who were unable to adopt a new view of their work have become alienated from it. It has proved difficult for some managers to understand the source of this alienation; that for the worker, the fundamental nature of work has been transformed. (Zuboff, 1988)

So we are using the techniques of oral history and video documentaries to record the responses of the coalition to the CSW. We are attempting to capture and preserve the human voices of the Washington effort. (Brunet et al., 1991)

The electronic studio itself offers an interesting way to evaluate the impact of this project. As an information repository, the studio embodies important historical records of the activities of the software engineers and the case management groups. This studio material can be made available as part of the final report of the project.

But, it would not be necessary for reviewers to wait for the termination of the project to review the studios. With current networking, they could access much of the information in the electronic studio as it is generated.

The electronic studio as an Internet collaboratory becomes a living repository that provides different views of our work to match different interests of outside reviewers.

THE COLLABORATORY IN THE NII

Defense researchers and basic scientists have taken the lead in constructing the initial networking and computing technologies required to support the collaboratories of the future. The recent expansion of the Internet and attention to the national information infrastructure have combined to motivate service providers, especially those in university settings, to further explore ways to exploit these technologies for their own distributed research and service delivery purposes. However, perhaps the greatest impetus propelling us toward distributed health care teamwork is the health care reform effort. The call for a national health information infrastructure has been one of the most bipartisan aspects of the entire health care reform debate (Silva, 1994). There is almost universal consensus at the federal level that the reform effort should be supported by a secure, nationwide information network.

This shared vision of how a reformed system of care can function in a distributed teamwork environment is based upon a model of patient centered care, similar to the one described by Barry Zallen, this volume, ch. 2). In a networked environment each patient could participate in information sharing with a team of providers through a card much like an ATM card, a point-of-service information system that collects information.

The system would automatically provide relevant patient information directly to all provider team members as a by-product of care. Medical knowledge, national outcomes data, and provider-specific results could be brought to any provider team member during their encounter with the patient. Uniform standards for administrative, clinical, financial, unique identifiers and other health care data could be developed for all health information used by all team members. Computerized patient records can be captured as a by-product of the delivery of care. The pharmacist, lab technicians, and other provider team members can contribute to the record and to health outcomes research aggregated.

However, there are several steps that must be taken before health care can be delivered effectively by a team of providers distributed across time and space. The following set of recommendations point to some of the first set of steps.

Standards must be developed that ensure the protection of privacy, security and confidentiality of all patients.

Computer hardware and software architecture should be developed that fully supports provider workstations and networks. Existing and emerging tools and standards for developing distributed component environments should be evaluated for their suitability to the health information industry and workstations in particular. The human-computer interface should be augmented by robust knowledge-based techniques that support context relevant presentation by improved voice-recognition and pen-based technologies.

Scenarios, modeling tools, evaluation criteria, requirements, and emerging standards should be shared throughout the health and human services community. The aim should be to create a comprehensive model of the information services needed in the professional environments—nursing, pharmaceutical industry, health and social service. Alternative methods for modeling and developing functional requirements should be formally evaluated through demonstrations, prototypes, and testbed applications. The criteria for evaluating those approaches should be developed through open, international efforts and be made available to all interested parties.

Criteria-based evaluation of domain models and reference architectures must be established to refine and evolve the health professional workstation. New evaluation strategies and metrics will be needed to increase the usefulness of the human interface.

REFERENCES

Brunet, L. W., Morrissey, C. T., & Gorry, G. A. (1991). Oral history and information technology: Human voices of assessment. *The Journal of Organizational Computing, 1*(3), 251–274.

DeGregori, T. (1985). *A theory of technology: continuity and change in human development.* Ames, OH: The Iowa State University Press.

Englebart, D. C., & English, W. K. (1968). A research center for augmenting human intellect. *Proceedings of the Fall Joint Computer Conference* (pp. 395–410), San Francisco.

Ferguson, C. H. (July–August, 1990). Computers and the coming of the U.S. Keiretsu. *Harvard Business Review, 68*(4), 55-70.

Gorry, G. A. , Long, K., Burger, A. M., Jung, C. P., & Meyer, B. D. (1991). The Virtual Notebook System: An architecture for collaborative work. *The Journal of Organizational Computing, 1*(3), 233–250.

Mintzberg, H. (1989). *Mintzberg on management: Inside our strange world of organizations.* New York: Free Press.

Nadler, G., & Hibino, S. (1990). *Breakthrough thinking.* Rocklin, CA: Prima.

National Research Council (1993). *National collaboratories: Applying information technology for scientific research.* Washington, DC: National Academy Press.

National Research Council. (1994). Appendix A. Federal networking: The path to the internet. *Realizing the information future: The internet and beyond* (pp. 237–253). Washington, DC: National Academy Press.

Schrage, M. (1990). *Shared Minds.* New York: Random House.

Silva, J. (1994, July 20). Testimony before the subcommittee on oversight and investigation, senate committee on veterans' affairs.

Wulf, W. (August 1993). *Science, 261,* 13.

Zuboff, S. (1988). *In the age of the smart machine: The future of work and power.* New York: Basic Books.

HEALTH INFORMATION

6 Shared Decision Making and Multimedia

John Wennberg
Dartmouth Medical School

If outcomes research is medicine's third revolution, as Arnold Relman (1988) suggested, then shared decision making must surely be its fourth. Outcomes research is the systematic recording of the collective experiences of representative doctors and patients to improve both our knowledge of the probabilities of outcomes contingent on alternative clinical strategies, and our understanding of how patients value these outcomes. But the potential of outcomes research to transform the health care economy, making it more efficient and humane, will not be met until doctors, patients and policy makers fully understand that health care decisions depend as much on personal values as on the probabilities derived from the clinical sciences.

The basis of medicine's fourth revolution is a shift from a reductionist, pathophysiologic model in which the role of the physician is to "discover and defeat" disease, to a model that places disease and its treatment within the context of the life of the individual. Modern health care technology and biomedical science have created so many opportunities to discover diseases, and so many options to treat them, that the preferences of the individual patient need even more consideration than in the past if we are to avoid depersonalizing health care.

The study of practice variations[1] and the conduct of outcomes research

---

[1]Practice variations are patterns of *practice style*—that is, the typical patterns in which physicians use hospital and other resources in treating patients—which drive the rates of local utilization of medical care. The earliest identified practice variations were in rates of tonsillectomy in the United Kingdom in the 1930s, which varied by up to tenfold between demographically similar communities. Similar variations in surgical procedures, hospitalization rates, and per capita expenditures on medical interventions have been identified throughout the United States.

strongly suggests that the irrational variations in practice patterns that characterize the current system are the result of supplier-induced demand. It has been shown that the per capita supply of hospital beds in a community drives the rate at which patients are hospitalized for conditions which can either be treated in the hospital or in another setting. The phenomenon of supplier-induced demand, which can be explained in part by uneven geographic distribution of specialists and variable untested theories of treatment effectiveness, also makes clear that provider preferences dominate the clinical decision making process. This is because, for many common conditions, locally held theories about the benefits of medical interventions, and the local supply of resources, rather than patient choice, determine the rate at which medical resources will be deployed. A remedy for the entanglement of preferences which leads to more—or at least different—care being prescribed for patients than they would choose for themselves, is the creation of systems which allow shared decision making between physicians and patients (Wennberg, 1992).

The choice among opportunities to diagnose and options to treat should be based on the patient's own preferences, which can only be known to the patient on the basis of his or her own life experiences. Not all patients even want to know they have a disease, particularly when there are no effective treatments; and for most conditions and diseases, there is no single treatment or single outcome that all patients want. Different treatments for the same condition have different risks and benefits, and individual patients value these outcomes differently. Shared decision making emphasizes communication, and the need to inform in ways that allow patients to understand their critical role in the decision.[2] It emphasizes the need to free the doctor-patient relationship from economic and other incentives that make it difficult or impossible for suppliers to provide information that is perceived by patients and payers as balanced, and presented in a way that helps patients make choices based on their own needs and wants.

Medicine's fourth revolution thus places unprecedented emphasis on the need for fair communication, the need to inform patients in an unbiased way about medical complexity, and the need to keep information about options and their possible outcomes up-to-date. The changes now occurring in

---

[2]Shared decision making is different from informed consent. Informed consent, which has been in effect in the United States for more than 20 years, is a legal document to verify that patients consent to the treatment their physicians have prescribed; in states that adopt the reasonable patient standard for informed consent, the doctrine could be viewed as transitional to shared decision making. It does not, however, remedy supplier-induced demand as evidenced by the persistence of practice variations. For example, informed consent is executed with care in Boston and New Haven, yet the risk for surgery varies substantially depending on which city patients live in and thus the physicians from whom they seek prescription of care. (Wennberg, Freeman, Shelton, & Bubolz, 1989)

information technology, particularly the so-called information highway and its opportunity for multi-media representations of medical choices, hold great promise. But a successful marriage between shared decision making and multimedia networks depends on new institutions and relationships among public, academic and private sector interests. We need credible ways to adjudicate and warrant the objectivity of messages that claim to fairly represent medical choices.

THE DILEMMA OF CHOICE ILLUSTRATED

Three examples from our research experience illustrate the importance of shared decision making.

Benign prostatic hyperplasia (BPH) is a common condition affecting many men after they reach middle age. Two widely held theories about treating BPH with surgery—that early treatment of the condition would increase the patient's life expectancy, and that surgery which improved urine flow would improve the patient's quality of life—have traditionally guided physicians' recommendations to their patients about how to treat the condition. But outcomes research has shown that the "best" treatment depends on the patient's own goals and concerns. Surgery for BPH has significant risk of both impotence and incontinence. If the patient is very bothered by his symptoms, not very concerned about sexual function, and not particularly averse to the risk of surgical complications, then prostatectomy is probably best for him, because it is likely to be effective in relieving the symptoms that bother him. On the other hand, if he worries about sexual dysfunction or is not all that bothered by his symptoms, then conservative treatment is probably better. Clearly, the choice of treatment ought to reflect the patient's attitudes about the risks of adverse outcomes and the relief of his symptoms, rather than the physician's.

Early stage cancer of the prostate presents another set of difficult problems of choice. Reliable information about the "main effect" outcome of treatment for early stage prostate cancer is not available. In contrast to many other conditions, clinical trials have not answered the question whether aggressive therapy (surgery or radiation treatment) improves life expectancy over watchful waiting.[3] It is well established, however, that cancer of the prostate is very common and, in most men, slow growing. More than 40% of men 70 years of age have microscopic cancers; for most men, the tumor is so slow growing that the patient will die from some other cause before the cancer causes any problems. The risks of aggressive treatment of

---

[3]Clinical trials designed to detect a 20% improvement in life-expectancy for aggressive treatment will not be completed for at least 12 years.

early stage prostatic cancer are well documented and substantial. They include impotence (affecting most men who have surgery, and a significant proportion who have radiation), incontinence, and treatment-induced mortality.

Men must decide whether the possibility of benefit, achieved some time in the future, is worth the costs of the up-front, immediate risks of impotence, incontinence, and treatment-induced mortality. Men who value their present quality of life or who are not persuaded by the argument for aggressive treatment may more often choose watchful waiting. Those who want to maximize the possibilities for longevity may want aggressive treatment.

The third example illustrates the dysutility of information. In the United States, men over 50 have been urged by some authorities to have a blood test measuring prostate specific antigen (PSA) as a screening test for early stage cancer of the prostate. But we have found that some men, when they are informed that the outcomes of treatment are uncertain—and when their own preferences are to avoid the risks of surgery or radiation—would rather not be screened at all. They do not want to know that they have a cancer for which they would choose not to be treated; and they do not want to have the status of a "cancer survivor." Not all men feel this way, but our experience is that a significant minority, when given enough information to make an informed choice, will choose not to be screened. For these men, the choice is not between treatments but whether or not they want certain information.

In patients with BPH, early stage cancer of the prostate, and other conditions, there are a range of choices that informed patients can make. How will they choose? We can only learn by asking. When doctors decide for them, we never know what patients want or how they would choose. Rationality in medicine depends on asking patients, because only patients can know and evaluate their own life situations. Choice cannot be rationally diagnosed by physicians on the basis of information available from the pathophysiologic database. Clinical histories, laboratory tests, physical examinations, even symptom scores do not provide the information necessary to prescribe what patients want. The information gained from screening tests impacts the lives of patients who are tested; whether or not to have the information should be an informed choice. When preferences are the seat of rational choice, then the information needed to make choices is found in the subjectivity of the individual.

BUILDING SHARED DECISION MAKING

Shared decision making depends on two complementary undertakings: understanding what is at stake in the decision, and representing the problem of choice in ways that empower the patient to choose. Shared decision making thus depends on outcomes research as well as on representations of medical choice.

Outcomes Research

The strategies for outcomes research have been presented in other contexts and will not be discussed extensively here. Briefly, outcomes research for conditions such as BPH or early stage cancer of the prostate is an ongoing, interactive strategy designed to keep abreast of new technologies, to focus on all of the outcomes that matter to patients, to achieve standardized instruments for measuring those outcomes and to provide reasonable estimates for the probabilities that the outcomes will occur, according to the patient's condition and the treatment used. The methods of outcomes research are eclectic, ranging from observational studies to clinical trials conducted on the basis of either randomization or patient preference for treatment. The development of new information media, in particular multimedia digital technologies using centralized servers, provides important new opportunities to undertake prospective clinical studies based on the preference study design.

Representations of Medical Choice

The communication tasks necessary for meaningful involvement of a patient in the decision making process are difficult. The doctor is a source not only of information about the probability of important outcomes, but also of vicarious experience of earlier patients whose lives were transformed by the same condition and its treatment. Approaches to negotiated decision making between doctor and patient have received a good deal of attention, but recognition of the importance of patient subjectivity raises novel and difficult questions. How can patients be empowered to make choices based on their own attitudes and values? How can the messages that empower choice be made fair and objective?

EMPOWERING PATIENTS TO CHOOSE

Shared decision making requires that patients understand what is at stake in choosing to diagnose or treat disease, and that they comprehend the predicament of choice. Patients must learn what works (when that is known) and share in the uncertainties (caused by weaknesses in the scientific basis of clinical decision making) when it is not. They must look ahead to assess what choosing the available alternative treatments would mean. Patients need to be empowered to understand that they really do have choices, and to learn the basic facts about possible outcomes and the chances of those outcomes

FIG. 6.1.   The videodisc player marries the computer to video to produce interactive multimedia with full motion video; we call the device the shared decision-making program (SDP).

occurring—situations and events that are abstractions to the personal experience of all but a few.

Some patients, when given access to detailed specific information about the future, choose to learn less. Systems to support shared decision making must, like the sensitive clinician, be able to accommodate different patients' different preferences for being informed at a particular point in time.

Our research group has been exploring ways to give people the information they need to make decisions. Sometimes, a simple pamphlet will work; but in cases where patients need to understand the structure of a difficult decision problem, and the tradeoffs associated with the multiple risks and benefits, written material alone may not be the optimal approach.

For BPH and early stage cancer patients, we are presenting the results of our outcomes research using the interactive videodisc player. The videodisc player marries the computer to video to produce interactive multimedia with full motion video. We call the device the shared decision making program (SDP, see Fig. 6.1).

The video provides a solution to the problem of how to convey pictures of the future that can help patients understand their predicament by presenting, on videotape, patients who have already been there. The SDP also helps convey probability information about outcomes, tailored to the individual. Using age, general health status, symptom level and history of acute retention, we classified BPH patients into multiple subgroups. For each of these subgroups, the probabilities for significant outcomes differed in important ways. For example, the chance of operative death is highly dependent on the patient's age. It would be misleading to quote the mortality rate for a healthy 65 year old if the patient making the decision was in poor health and 85 years of age. It would also be wrong to quote the chances for symptom relief without being specific about the changes in symptom status the patient could experience.

The computer can easily calculate the patient-specific probabilities from the database and then display them graphically, as long as the SDP is used in a clinical situation where the provider can supply the needed information on the patient's clinical background (see Fig. 6.2). (In the case of PSA screening we used a linear video approach because the individual deciding on a test is not necessarily in a clinical setting.)

The SDP is designed to be used in everyday practice to support shared decision making. The first task is to set the stage for the shared decision making model. In the PSA video, the stage is set by Dr. C. Everett Koop:

> I think it's very important that men see the following video on prostate cancer screening. It's a much bigger story than "just a simple blood test," and it's representative of many other issues we all face in our health care decision

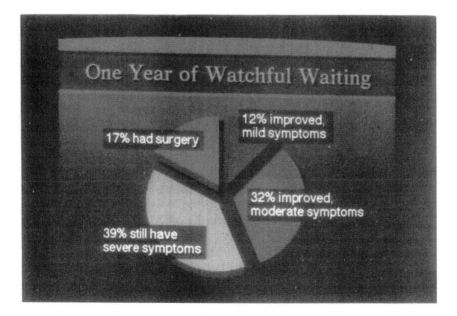

FIG. 6.2.   The patient is given information specific to his own clinical status; in this case, the outcomes that are likely to happen to a man with severe symptoms if they choose watchful waiting.

making. We want to help you get the facts so you can get involved in your own health care decisions—as a partner with your doctor in what we call "shared decision making." There's a lot changing in medicine today but one thing hasn't changed: What *you* think is important is important. We hope this video will help you think about your values, so you can make the decision that's best for you. ("PSA Screening," 1994)

For the SDPs on PSA and BPH, the heuristic we chose to empower patients to understand that they really do have a choice was to present interviews with two physician-patients, each of whom had experienced severe symptoms, but who each chose a different treatment strategy. Our logic for selecting physicians was that if patients see that physicians can choose differently, they understand that so can they.

Dr. A, who chose watchful waiting, explains to the viewer his approach to risk assessment:

I considered the advantage of the operation against the amount of trouble I am having with the symptoms and the extent to which [the operation] might relieve them. And I felt that I am not bothered enough even by these fairly severe symptoms to undertake the risk of incontinence which the operation involves. ("Benign Prostatic Hyperplasia," 1994)

Dr. B, on the other hand, emphasizes the amount of trouble he was having with the symptoms and how they interfered with the quality of his life:

[It was] the feeling that I had a full bladder, to know that it took a long time to empty it . . . and the fact that I would have to wake up more often at night. And again, the restrictive features, to be able to do less and less things or to worry about more and more things as I began to plan my daily routine. ("Benign Prostatic Hyperplasia," 1994)

The two physician patients have typical outcomes following surgery and watchful waiting. Dr. B is among the 80% of men who have a very satisfactory result from surgery. The narrator asks him how he fared. "Oh, infinitely better. Just a totally different situation. Such a feeling of relief. I remember the day when I walked into my urologist's office, and [begins singing]: 'Summertime, and the peeing is easy'. That's the way I felt" ("Benign Prostatic Hyperplasia, 1994).

Dr. A, who is severely affected by BPH, has symptoms that are worse than many BPH patients who watch the video. His description of his lifestyle as a watchful waiter offers viewers insight into the situation they may face if they choose not to have surgery now. "I've made a lot of adjustments and it has taken a lot of planning and anticipation. For example, all tickets on the airliner or concerts or theaters had better be on the aisle side so I can get out in [a] hurry if I need to. I don't go through a three hour movie without having to leave" ("Benign Prostatic Hyperplasia," 1994).

Like the BPH program, the SDP on PSA presents two physicians, each about 65 years of age; one has an annual PSA, the other does not want to be tested. The goal is to let the patient know there is a good deal of uncertainty about the value of aggressive treatment and therefore the value of the test, and that informed patient-physicians who know about the uncertainties can make different choices.

SDP programs also present interviews with patients who have experienced complications. The BPH-SDP, for example, shows one patient's experience with watchful waiting, and one patient's experience with surgery. Each experience has about the same chance of occurring.

A patient with surgically-induced incontinence reports:

> I have leakage. I think the word is incontinence or something like that. And I was getting wet all the time and of course I didn't know what to do about all that so my wife and I figured it out. I went out and bought some jockey shorts and sort of Kotex type stuff and put it inside and I would have to change that three or four times a day, which I am still doing. It didn't pour out, but it would on occasion leak out and was much worse when I walked around a lot or stood a lot . . . This came out of the blue [following surgery]. This was a minus, a big minus. ("Benign Prostatic Hyperplasia," 1994)

A watchful waiter who had an episode of acute retention answers questions about his experience.

> Q. Were you in a lot of pain?
>
> A. Yes, a lot of pain; pain that I couldn't control or help. So finally, in getting up to the doctor, I got on the table to be interviewed there and I said, "Hold on, doctor, before you go any further, the first thing you do is drain that bladder."
>
> Q. How did it feel when they finally did use the catheter?
>
> A. Heavenly! It was like being under water longer than you wanted to be and you had to hold your breath longer than you wanted to and the moment that catheter cleared the passage there, it was a relief like that pain was leaving all the time—right up until it got comfortable—the doctor made two or three trips with the urinal bowl until there was no more. ("Benign Prostatic Hyperplasia," 1994)

Do patients really want to know about the options for treatment? The BPH-SDP, which has been in use the longest, has received the most extensive evaluation. By the summer of 1992, over 1,000 patients had viewed the program.[4] We asked patients, "In general, how do you feel about patients seeing a presentation like this before deciding whether or not to have prostate surgery?" Seventy-seven percent chose "very positive" as their response, whereas another 16% said they were "generally positive;" 6% chose "neutral," and only 1% chose "somewhat negative." Although those who had received less formal education seemed a little less enthusiastic, 69% of those who didn't finish high school indicated they were "very positive," and 92% were "generally positive" or "very positive."

The most important point, however, is that patients actually do become empowered to choose their own treatments. Only a small minority of patients were unprepared or unwilling to participate in shared decision making.

FAIRNESS, OBJECTIVITY AND THE AUSPICES FOR REPRESENTING MEDICAL CHOICES

The fourth revolution brings an important new challenge. Is it possible to assure the balance, fairness, and authenticity of information that will justify the claim that shared decision making is a better way to communicate options than the traditional strategies? How can we choose between alternative ways of representing medical choice in the shared decision making environment? These are difficult scientific, intellectual, and philosophical issues. Because we recognize their importance, we have taken some pragmatic steps.

Constructing Representations of Choice

The BPH-SDP project began as an experiment to learn if we could effectively communicate the complexity of the BPH decision to patients and, if so, if patients would be willing to adopt a new role in making medical choices. As it became clear that patients wanted to assume responsibility for the choice of treatment, the question of how to construct and maintain such representations became increasingly important to our group. We set as our goal the organization of SDPs to cover the major treatment dilemmas for which patient preferences are critical. Table 6.1 sets out the conditions that

---

[4]Most (87%) agreed that the information presented was the amount they needed to make their decisions, whereas 6% said they would like more and 7% felt they learned more than they needed to know. The rating of the length of the program was equally gratifying, with only 5% thinking it too long. Ninety-nine % thought that everything or most things were clear. Only 5% of high school graduates, and 16% of men with less than a high school education, felt that they received *more* information than they wanted.

TABLE 6.1
Conditions and Treatment Options Covered by Shared Decision Making
Programs

Condition	Treatment Options
Stable angina (pain due to obstruction of arteries of the heart	Bypass surgery versus drug treatment
Mild hypertension	Drug therapy versus life style modification versus watchful waiting
Benign prostatic hypertrophy (enlarged prostate)	Prostatectomy versus balloon dilation versus various drugs versus watchful waiting
Breast cancer I	Mastectomy versus breast conserving surgery followed by radiation
Breast cancer II	Adjuvant chemotherapy and/or adjuvant therapy or no adjuvant therapy
End of life care	Pros and cons of living will/advanced directives with strategies for adoption
Low back pain	Surgery versus medical management
Benign uterine conditions	Hysterectomy versus hormone replacement versus watchful waiting
Menopause	Hormone replacement therapy or watchful waiting
PSA screening test	Active screening versus watchful waiting
Prostate cancer (early stage)	Radical prostatectomy versus radiation versus watchful waiting

we consider to have high priority. To accomplish our goal for these and other common medical conditions, we needed an organization that would be viewed as independent of interests vested in particular solutions to the problems of choice represented by the SDPs. For example, an SDP on prostate disease financed by a drug company that had invested in the PSA screening test, or in drug treatments for BPH, or by device manufacturers, or by the American Urological Association, could not make a successful claim to objectivity or balance, no matter how well the processes to establish the claim might be conducted.

To produce the SDPs independently, we established the Foundation for Informed Medical Decision Making, a not-for-profit organization. The Foundation's mission is to:

1. Organize reliable and continuous access to the outcomes research community to assure that the scientific representations embedded in the SDPs are up-to-date and as accurate as they can be.

2. Organize a process for producing and periodically updating the SDPs that ensures that they are fair and balanced.
3. Promote widespread research into the effectiveness of SDPs and alternative ways of achieving shared decision making.
4. Improve the scientific basis of medicine by promoting prospective clinical trials based on preference as well as randomized designs.

The foundation has formed links with the outcomes research community to create SDPs for nine conditions (see Table 6.1).

Balance in the Design of SDPs

There are no firm scientific paradigms or ethical principles that prescribe the rules and procedures the foundation should follow to assure balance and fairness. The principal goal is to assure that the outcomes that matter to patients are clearly and fairly represented in the script and in the videotapes of patients. Our principal means of assuring compliance with our goals of balance and fairness are collaboration with outcomes researchers and the foundation's conducting of a series of focus groups.

We use focus groups to ascertain what outcomes matter to patients, so that concerns about those outcomes can be explicitly addressed in the SDP. Patient focus groups review the SDPs to assure that their expectations and concerns are adequately addressed. Focus groups with physicians and other interested providers help negotiate the disagreements about balance that naturally emerge from those whose interests and preferences for treatment differ by virtue of their specialty, financial interests or personal convictions.

Patient Focus Groups.   In the absence of formal procedures, we elected the pragmatic approach of structured and unstructured interviews with patients. We ask groups of patients to explain their perceptions about their conditions, the symptoms that bother them, and what it is they want to know in choosing between the available treatments. For those with complications, we ask what they would want to have known about possible complications before they made their decisions. Patients' fears, expectations, and misunderstandings are identified, and the information is used to plan the presentation so that it can correct significant misperceptions. In the case of BPH, for example, we found that some men were not informed about retrograde ejaculation before surgery. We also learned that many men thought having a prostatectomy would reduce their chances of getting prostate cancer, a completely wrong perception that led to wrong decisions. To deal with these misperceptions, we added sections to the program that specifically address these issues.

Focus Groups with Providers.   The SDPs are reviewed by physicians and other providers who have a stake in the decision. We have observed that physicians seem to systematically believe that the SDP is biased in favor of a specialty other than their own; for example, even after several focus sessions with urologists and internists, in which changes to improve balance were negotiated jointly, each group of specialists remained convinced that the SDP had a slight bias in favor of the other group's treatment preference. The process, however, with its emphasis on negotiation and discourse, has the happy result that each side sees the source of bias not in the SDP itself but in their differing points of view. Because the providers in our focus groups are committed to shared decision making, they recognize that the final arbiter of balance is the patient.

The Objectivity of Shared Decision Making

If rational choice depends on the subjective response of the patient, how can we obtain objective information that decisions made under shared decision making are better than those made under the old paradigm of leaving choice up to the provider? The concept itself, including the notion that SDPs can be iteratively improved, depends on a research agenda that is only now beginning to be organized as part of regular discourse among evaluative scientists. In the case of the evaluation of BPH-SDP, some important and encouraging facts have emerged.

Under shared decision making, the profile of patients choosing surgery is different than the profile of those who were prescribed surgery. Most men as severely symptomatic as Dr. A would have been prescribed surgery by their urologists. Under shared decision making, patients who were severely symptomatic were about twice as likely as moderately symptomatic men to choose the treatment with the best chance of improving symptoms, namely surgery. But across the spectrum of symptom severity, even among the most severely symptomatic, only a minority of patients wanted surgery. Less than 1% of those with mild, 11% of those with moderate, and 22% of those with severe symptoms chose surgery.

The most important finding, however, is that *when decisions among treatments for BPH are made according to the shared decision model, it is how the patient evaluates his own attitudes toward risk and his symptoms that matters most in determining choice.* How do these two spheres—the objective state of symptom level and the subjective attitude of patients about them—interact to predict the choice that patients will make in the shared decision making environment? Drs. Fowler and Barry, the members of our research team who undertook this investigation, asked patients to rank the degree to which they were bothered by their symptoms and their attitudes about impotence. They then studied the relative importance of these factors and the objective level of the patients' actual symptoms in the decision to choose surgery or watchful waiting.[5]

When patient attitudes about the possibility of impotence and the degree to which patients were bothered by their symptoms were taken into account, *symptom level per se no longer predicted the choice of treatment.* In other words, SDP helped patients make choices that reflected their subjective attitudes about their symptoms and their concerns about risk. What matters to patients is their degree of botheredness with urinary dysfunction and their concerns about impotence. Patients who were negative about their symptoms were seven times more likely to choose surgery than those who had a positive or a mixed attitude; those who were negative about the prospect of impotence were five times more likely to choose watchful waiting than those who had mixed feelings or didn't seem to care.

The opportunity to obtain information on subjective states—and to correlate the information with actual choices—provides a framework for validating and advancing the knowledge base for preference-based decision making. Through our initial studies in Maine we learned that many men with BPH who were risk averse or who were not bothered all that much by their symptoms were nonetheless prescribed surgery, because their physicians believed that symptom relief was the most important issue. We were gratified to see that our first edition SDP remedied this problem; but a good deal of the variation in decision making remains unexplained. This suggests that other, unmeasured, subjective factors are playing a role. Improvements in the SDP may be possible when these factors are measured and related to decision making. SDP provides a standardized approach to informing patients; its messages can be altered in experiments to test framing effects and other features of communication that may influence choice. It thus provides the basis for its own iterative improvement, and in this respect, we believe the SDPs can contribute to basic clinical knowledge. We also hope to help

---

[5]Barry and Fowler used the following question to ask patients to rank their attitudes toward their symptoms: *"Suppose your urinary symptoms stayed just about the same now for the rest of your life. How would you feel about that?"* In answering this question, patients were asked to check one of five categorical responses: delighted, pleased, mostly satisfied, mixed, mostly dissatisfied, unhappy, terrible. Based on these responses, patients were grouped into three classes: those who were positive (the first three responses), mixed, and negative (the last three categories).

Surprisingly, most were satisfied by the prospect that their symptoms would remain the same. Even among those who were severely symptomatic: 46% were satisfied, 21% had a mixed reaction and only 33% expressed negative feelings about continuing in their current state. Among those with moderate symptoms, the ratings were 62% positive, 19% mixed and 19% negative; among those with mild symptoms, 85% were positive, 12% mixed and only 3% negative.

A similar set of questions was asked about impotence.

patients anticipate how a particular condition and its alternative treatments will transform life by measuring anticipated utility for specific states prior to the treatment decision and then measuring quality of life when those states are reached.[6]

SHARED DECISION MAKING AND MANAGED CARE

The prospects for shared decision making depend on the willingness and capacity of health care organizations and providers to yield responsibility for choice to their patients for those crucial decisions where rationality depends utterly on patient preference. This does not come easily. Sharing decision making with patients requires providers to rethink what for many are deeply ingrained attitudes based on the time-honored authoritarian role of the professional. Even more problematic are the economic disruption and the chaos that patient-centered choice introduces. Shared decision making disrupts the dysequilibria between supply and utilization. Revenues are no longer predictable, as the market seeks a new equilibrium based on "true demand."

The classic health maintenance organizations (C-HMO) such as Kaiser Permanente and Group Health Cooperative of Puget Sound, provide an organizational and financial structure and a practice culture conducive to shared decision-making. Because physicians are salaried, personal economic incentives do not become entangled with medical choices. For C-HMO patients with angina, economics do not drive surgeons to perform operations, cardiologists to perform angioplasty, or primary physicians to prescribe drugs. Because the hospital does not depend on fee-for-service revenue to meet its budget, operating rooms do not need to be utilized or beds filled. Because all physicians work for the same firm, a culture of shared decision making can be articulated by leadership and supported by corporate policy, and the organization can adapt to new levels of utilization. Use of the BPH-SDP in the Kaiser Plan in Denver and the Group Health Plan in Seattle led to a 40–50% decline in the per capita rates of prostate surgery.

For C-HMOs, shared decision making can stand as an emblem of quality and an ethical imperative. The uncovering of true demand in this fashion can transform the relationship between payors, managed care companies, doctors, and patients. The assurance that demand comes from patients and not from providers is a categorical defense against the charge that HMOs undertreat their patients. Knowledge that for conditions such as BPH many

---

[6]Approaches to measuring how patients value the various possible outcomes of treatment decisions, and their predictive validity for future quality of life, have been described elsewhere.

plan members prefer a less expensive option can enrich the debate about entitlements and cost sharing.[7]

It is not clear whether the incentives operating in other forms of managed care will effectively support the implementation of shared decision making. In networks, physicians are often paid on a fee-for-service basis; hospitals usually depend on utilization to obtain the revenues to meet expenses. Networks do not characteristically own their own hospitals or employ their own specialists; they will therefore encounter difficulty in developing a culture in which the ethic of shared decision making is held by all relevant decision makers. Because the network form of managed care is expected to become the dominant model under health care reform, the prospects for shared decision making remain problematic. Whereas in some places in the United States networks may evolve to become C-HMOs, many communities do not have the population base needed to support them under a competitive model. In such regions some form of regional organization may be needed to support the culture and to build the incentives conducive to shared decision making (Wennberg & Keller, 1994).

Shared Decision Making and the Information Revolution

The revolution in communication technology, through the developing information highways, can facilitate the spread of shared decision making.

First, it can help with the logistics and the complexities of organizing such an information system. Whereas stand-alone CD-interactive or videodisc players accomplish the goal of tailoring communication to the needs of specific patients, the decentralized, stand-alone platform has important disadvantages. The SDP strategy depends on tight control over the update of information. Discs become outdated and must be revised as often as every 6 months. A centralized digital server would be much more easily, and more reliably, updated than the current system, which depends on periodically replacing videodiscs or CD-ROMs at individual practice sites. Moreover, the use of interactive media in real time for clinical decision making could be facilitated if the relevant information were also available in the patient's home. A centralized digital server connected to users is more efficient for this purpose.

The most important advantage of the marriage between shared decision making and the new information technology is the potential for converting

---

[7]The demonstration that choice is preference based and that patients with similar diseases and levels of severity choose differently, opens the possibility that more expensive options need not be fully subsidized. At the point-of-service, prostate surgery in Kaiser Permanente and Group Health Cooperative is obtained without cost. If a copayment had been introduced, the rates may well have dropped further.

everyday practice into a learning laboratory. The new digital highways for shared decision making, served by central servers, will be ideally organized to add to the scientific basis of clinical medicine. Take, as an example, the problem of the outcomes of aggressive treatment versus watchful waiting for early stage cancer of the prostate. Randomized clinical trials to obtain this information are just getting underway. Assuming that enough men are indifferent to the risks of surgery or radiation and are willing to accept randomization, it will still take years before the evidence from these studies is available. In the meantime, thanks to PSA, each year more than 100,000 American men are being treated for this condition: some with watchful waiting, others with more aggressive treatment. With few exceptions, the experiences of these men—in terms of the outcomes of their care—go unrecorded.

This is unnecessary. Shared decision-making and outcomes research can be linked through the conduct of preference trials, clinical studies similar to randomized clinical trials in which all patients are followed up according to the kind of treatment they receive. The single difference is that in the preference design, patients actively choose their treatments, and in randomized trials patients are assigned treatment on the basis of a flip of a coin.[8] Because every patient, on the basis of his or her own choice, becomes a candidate for a preference trial, the potential number of patients available is limited only by the number of incidence cases, the capacity to organize the studies, and the sophistication and ease of the mechanisms for follow-up.

The organization of preference trials around shared decision-making will be greatly facilitated by the central servers. The necessary clinical information needed to identify the patient's subgroup is entered at the time of viewing. Information on choice follows, and follow-up is greatly facilitated, particularly in a reformed health care system where electronic billing information makes it easy to locate patients for subsequent follow-up.

CONCLUSION

Shared decision making is the logical next step for the outcomes research movement. Rational choice among options depends on patient preferences. This can only be ascertained by asking patients what they want, after they have understood the complexities of the available options. The Foundation for Informed Medical Decision Making has been organized to develop strategies to represent medical choices to patients, based on the results of

---

[8]The idea of a preference trial and a proposed series of methodologic studies to distinguish its advantages and disadvantages from classical randomized clinical trials have been described elsewhere. (Wennberg, 1992).

outcomes research. Studies of the foundation's products demonstrate that clinical decision making can be based on shared decision making, and that decisions made this way are more rational *from the patient's point of view.* For shared decision making to work, information must be constantly updated, must fairly represent the problem of choice, and must be free from influences that discredit its objectivity.

REFERENCES

Benign prostatic hyerplasia: Choosing surgical or non-surgical treatment (SDP). (1992). (Available from Foundation for Informed Medical Decision Making, P.O. Box 5457, Hanover, NH 03755-5457).

PSA Screening [Videotape]. 1994. (Available from Foundation for Informed Medical Decision Making, P.O. Box 5457, Hanover, NH 03755-5457).

Relman, A. (1988). Assessment and accountability: The third revolution in medical care. *New England Journal of Medicine, 319,* 1220–1222.

Wennberg, J. (1992). Innovation and the policies of limits in a changing health care economy. In A. C. Gelijns (Ed.). *Medical innovations at the crossroads, Vol. 3. Modern methods of clinical investigation* (pp. 9–33). Washington, DC: National Academy Press.

Wennberg, J., & Keller, R. (1994, Spring). Regional professional foundations. *Health Affairs,* pp. 257–263.

Wennberg, J., Freeman, J., Shelton, R., & Bubolz, T. (1989). Hospital use and mortality among medicare beneficiaries in Boston and New Haven. *New England Journal of Medicine, 321,* 1168–1173.

Public Health Information and the New Media: A View from the Public Health Service

J. Michael McGinnis
Mary Jo Deering
Kevin Patrick
U.S. Public Health Service

Disease prevention and health promotion represent time-tested ways for bringing all Americans closer to their optimal health through individual and social efforts. The price may be cheap as in a personal decision to quit smoking, or low as when large populations receive the benefit of public education for good nutrition. Many of these efforts are information based; helping people acquire the facts, understanding, and skills that enable them to reduce their health risks or participate responsibly in managing their own health.

Emerging communication technologies offer potent tools for strengthening and extending these efforts. From on-line information resources to interactive multimedia educational materials and computer-assisted decision-support programs, the new media can contribute to a richly diverse health support system. Many elements of this system are likely to be provided by the private sector, or through public-private partnerships. The federal government must reassess its role in relation to a wider array of information producers and providers. The challenge will be to make sure the system benefits all Americans, and that its design is responsive to national opportunities to reduce health costs rather than only commercial opportunities.

> Every year, the Federal government spends billions of dollars collecting and processing information. Unfortunately, while much of this information is very valuable, many potential users either do not know that it exists or do not know how to access it. We are committed to using new computer and networking technology to make this information more available to the taxpayers who paid for it. (Clinton, 1993)

Let me be clear. I challenge you [the telecommunications industry], to connect all of our classrooms, all of our libraries, and all of our hospitals and clinics by the year 2000. We must do this to realize the full potential of information to education, to save lives, provide access to health care and lower medical costs. . . . The best way to do this is by working together. . . . We must build a new model of public-private cooperation that, if properly pursued, can obviate many governmental mandates. But make no mistake about it—one way or another, we will meet this goal. (Gore, 1994)

THE CHALLENGES FOR HEALTH IN AMERICA

Disparities in Health

The attention on reforming American health care is long overdue. In 1993, our national health care expenditures reached about $900 billion, or over $14,000 annually for each family of four. Not only is the absolute amount extraordinary, but the percentage of GDP allocated to health in this country far exceeds that of other industrialized nations. Other nations spend roughly 8 to 10% of their GDP on health; we currently spend over 14%. Despite recent reports about slower growth of health care costs, the rate of increase still exceeds the general inflation rate. If this expenditure were associated with markedly greater indices of either personal or public health, it might be considered more defensible. This is not the case, however. In fact, in many areas, such as infant mortality, our rates place us well below other nations that spend relatively less for health care. Moreover, lack of access to preventive, diagnostic, and other medical services contributes to a situation of health "have and have nots" that is inimicable to our national culture. Of particular concern is the fact that certain populations—minorities, people with low income, and the elderly—carry a disproportionate burden of health problems.

Equally disquieting is the fact that so many Americans succumb prematurely. The human and economic costs are reflected in the nearly 12.3 million potential years of productive life that were lost in 1991 as a result of deaths that occur before age 65. (U.S. Department of Health and Human Services [DHHS], Centers for Disease Control and Prevention, 1993.)

The Role of Prevention

There is heartening news as well: research shows that approximately half of all deaths in America can be attributed to factors that we, individually and as a society, can control. (See Table 7.1). And information-based activities are our primary tools.

The U.S. Preventive Services Task Force, charged with evaluating alter-

TABLE 7.1
The Five Leading Causes of Death in the United States
and Associated Risk Factors

Cause of Death	Risk Factors
Cardiovascular disease	Tobacco use
	Elevated serum cholesterol
	High blood pressure
	Obesity
	Diabetes
	Sedentary lifestyle
Cancer	Tobacco use
	Improper diet
	Alcohol
	Occupational/environmental exposure
Cerebrovascular disease	High blood pressure
	Tobacco use
	Elevated serum cholesterol
Unintentional injuries	Safety belt noncompliance
	Alcohol/substance abuse
	Reckless driving
	Occupational hazards
	Stress/fatigue
Chronic lung disease	Tobacco use
	Occupational/environmental exposures

Note. From U.S. Department of Health and Human Services, National Center for Health Statistics (1993). Reprinted by permission.

native means of reducing disease, disability, and premature death, concluded that

> primary prevention as it relates to such risk factors as smoking, physical inactivity, poor nutrition, and alcohol and other drug abuse holds generally greater promise for improving overall health than many secondary preventive measures such as routine screening for early disease. (U.S. Preventive Services Task Force, 1989)

Other research suggests that a full array of primary prevention measures such as early detection and intervention, changes in individual behavior, and healthy public policies relating to firearms, motor vehicles, and occupational and environmental hazards could prevent between 40 and 70% of all premature deaths, a third of all cases of acute disability, and two-thirds of all cases of chronic disability (McGinnis, 1988–1989). Preventive programs launched in the 1970s are credited with fostering declines of more than 50% in stroke

TABLE 7.2
Leading Causes of Death, 1990

Heart Disease	720,058
Cancer	505,322
Cerebrovascular Disease	144,088
Unintentional Injuries	91,983
Chronic Lung Disease	86,679
Pneumonia and Influenza	79,513
Diabetes	47,664
Suicide	30,906
Chronic Liver Disease/Cirrhosis	25,815
HIV Infection	25,188

Real Causes of Death, 1990

Tobacco	400,000
Diet/Activity Patterns	300,000
Alcohol	100,000
Microbial Agents	90,000
Toxic Agents	60,000
Firearms	35,000
Sexual Behavior	30,000
Motor Vehicles	25,000
Drug Use	20,000

Note. From U.S. Department of Health and Human Services, National Center for Health Statistics (1992—top) McGinnis and Foege (1993—bottom). Reprinted by permission.

deaths, 40% in coronary heart disease deaths, and 25% in overall death rates for children (U.S. DHHS, Public Health Service [PHS], 1991).

But there remains a troubling disparity between these prevention opportunities and the efforts we, as a nation, are making to achieve their full promise. The great preponderance of health care dollars are devoted to treatment of the conditions listed in Table 7.2 as *leading causes of death* rather than to controlling the factors identified as *real causes*. Our national investment in prevention is estimated at less than 5% of the total annual health care cost (McGinnis & Foege, 1993). We clearly must take a closer look at the appropriateness—and effectiveness—of this approach. Focusing greater efforts on the root determinants of death and disability makes sense both medically and economically. A mechanism for doing so is already in place.

Healthy People 2000: The Nation's Prevention Agenda

Healthy People 2000, National Disease Prevention and Health Promotion Objectives, was released in 1990 (U.S. DHHS, PHS, 1991). It is the product of an unprecedented national consensus effort led by the U.S. Public Health

Service. Whereas health care reform aims to improve the ways in which health care is delivered, evaluated and financed—in essence the what and how of the medical system, *Healthy People 2000* clarifies the what and why of that set of activities which promise the greatest results in improving health status. *Healthy People 2000* proposes three broad goals for America to attain by the year 2000: (a) increase the span of healthy life, (b) reduce the disparities of health status among population groups, and (c) to achieve access to preventive services for all. Although the third goal may be achieved in a single bound if preventive services are included in the basic benefits package covered under health care reform, many individual steps will be needed for us to reach the other two.

These steps are spelled out in 300 specific objectives in 21 different priority areas ranging from physical activity and fitness, to environmental health, to HIV infection. Completion of this national prevention agenda by the year 2000 could yield dramatic health gains, such as a 31% reduction in infant death rates, a 30% decrease in the rate of teenage pregnancies, a 26% reduction in coronary heart disease deaths, and a 12% decrease in alcohol-related motor vehicle crash deaths. (U.S. DHHS, PHS, 1991)

Disease Prevention and Health Promotion are Information Based

In reviewing 169 different health care interventions for preventing 60 different illnesses and conditions, the U.S. Preventive Services Task Force found that the majority of effective approaches involve patient education and counseling. (U.S. Preventive Services Task Force, 1989). Many of these clinic-based educational efforts focus on lifestyle issues, as do those delivered through mass media and community-based programs. Other disease prevention and health promotion activities seek to motivate behavior change and promote informed decisions about the appropriate use of preventive or treatment services. Historically, these information-based efforts have relied on print and broadcast media and interpersonal communication.

PUBLIC HEALTH COMMUNICATION: ACTIVITIES AND PRINCIPLES

Past and Present Public Health Communication Activities

Public health has long included information and education activities among its efforts to control disease and reduce death and disability. Around the turn of the century, local health departments tacked up fliers on "How to Keep a Safe Privy" and mounted sanitation education outreach campaigns.

As military service overseas heightened exposure to venereal diseases, factual prevention brochures were supplemented by graphically dramatic posters with clever slogans that have a familiar ring today, like "NO is the best tactic; the next PROphylactic!" (Mullan, 1989).

Beginning in the 1970s, the power of mass communication was enlisted in the fight against smoking and other public health hazards. By the end of the 1980s, sophisticated research projects had demonstrated that the most effective efforts were multifaceted, combining communication (including messages directed toward medical professionals) with community-based health education activities and appropriate supporting services (Farquhar et al., 1990). As a result, public education in the 1990s is often anchored in an array of local activities and resources targeting not only individual behavior, but also cultural norms and structural supports, like laws, regulations, and policies. More personalized health education is delivered in schools and clinical settings. Comprehensive programs in elementary, secondary, and postsecondary settings, and patient counseling provide opportunities to convey information and develop skills necessary to use the knowledge gained. In all these developments, there is increasing interest in participatory approaches that engage people in their own learning.

The past 20 years have also seen a massive knowledge transfer effort, intended to support behavioral decisions, appropriate self-care, and informed medical consumerism. Personal responsibility for health entails not only healthy lifestyle patterns but also knowledge about health problems, when and how to seek appropriate medical attention, and how to participate in the management of medical problems. There is now a rich biomedical and prevention research base on topics ranging from Alzheimer's disease and cancer to nutrition, physical fitness, and rare diseases. It is translated into health information for specific audiences, organized, packaged, and made available to individuals (and/or intermediaries) by numerous federal and private information services. By 1982, for example, the U.S. Department of Health and Human Services (mostly the Public Health Service), supported 33 clearinghouses, collectively employing over 400 workers with combined operating budgets of almost $19 million. (U.S. DHHS, Office of the Assistant Secretary for Public Affairs, 1985.) By 1993, the number of information centers operated under contract had remained relatively stable but the total number of health information resources supported by or within federal agencies had risen to 123.

Requests for health information services have risen dramatically in the 1990s. Partial, preliminary data show that telephone inquiries increased an average of 209% since 1990, and mail inquiries grew by 43%. (Deering, 1994). Much of the growth of telephone inquiries reflects the availability of toll-free numbers. Automated response systems are permitting more efficient service. These health information services are appreciated well beyond the private circles of individual end users.

It has become clear that the most cost-effective thing government does is disseminate health information. (Will, 1989)

Other Health Information Resources

Borrowing an economic perspective, we can observe that the burgeoning demand for health information has spawned a tremendous "market" for this "commodity." The federal government is only now one producer and distributor in this field. Voluntary health agencies (VHAs), such as the American Cancer Society, are major information providers. Forty of the largest VHAs reported spending over $623 million on health education and information in 1991. Just 11 of them reach 82 million potential information consumers and answer over 4 million direct inquiries each year (Harris, 1994). Public libraries are a frontline resource for health information inquiries. They report that up to 10% of all reference questions are health related, which translates into about 52 million direct inquiries each year (Harris, 1994). In efforts to control health care costs, employers have become important sources of health and wellness information. The 1992 national survey of worksite health promotion found that 81% of private worksites with 50 or more employees offered health promotion activities, including information and resource materials about specific risk factors and health problems. This is up from 66% in 1985 (U.S. DHHS, PHS, 1993). For the same reason, many insurance companies, HMOs, and other health care providers are also offering information services for subscribers and patients. Health periodicals, computerized information services, foundations, research centers, medical and allied health schools, and commercial publishers are other key producers of health information. The sheer variety of sources, packagers, and distributors underscores the need for a reappraisal of the federal role. It also points to the twin problems of information overload and fragmented information sources that confront people who need answers to their health questions.

Challenge: Patterns of Information Demand are Changing

Public information requests have always covered a range of health issues, such as wellness and prevention, specific disease symptoms and treatments, prescription drug use, and support services or self-help resources for various conditions. There is some indication that requests for general wellness information may have been strongest in the earlier days of the consumer health information movement. For example, a 1985 survey of DHHS clear-inghouses identified "management of personal life style" and "upbringing of children and education of other family members" as the primary informa-tion uses of the general public (National Capitol Systems, 1986). More recent reports suggest that the public is becoming more interested in manag-ing personal health care and is seeking more information about specific diseases, drugs, and medical care treatment and financing options (Rees, 1991). As findings from health care outcomes research begins to reach the public, there appears to be growing interest in issues related to costs, quality, appropriateness, and effectiveness of medical care. Clearly, we must build greater medical literacy, while expanding general health literacy, if we are to develop consumer competence and responsibility.

"The present picture of consumer health information delivery is charac-terized by fragmentation rather than cooperation" (Fecher, 1985). The sheer number of health information sources indicates the wealth of this collective resource, but also highlights the daunting task facing an individual trying to find the answer to a specific health question. "According to Finagle's law, the information you have is not that which you want, the information that you want is not that which you need, and the information you need is not that which you can obtain" (Ashton, 1992). Even sophisticated informa-tion users must navigate difficult print, electronic, or telecommunications approaches to find the right path. Their experience often confirms that this path leads through somebody, a knowledgeable person able to help them negotiate the health information terrain. Thus, achieving an equitably in-formed citizenry will require addressing this potential deluge of disparate, unsynthesized information. We need better translation, management, pack-aging, and targeted dissemination to ensure that health information is timely, usable, and appropriate for different users' needs and learning styles.

Federal public education efforts face different access challenges. Compe-tition for mass media attention and "message clutter" make it difficult to reach effectively the general population with public service announcements. The resources involved in mounting a multifaceted communication cam-paign, adequately grounded in community activities, require well funded programs. As a result, there are relatively few comprehensive health com-munication campaigns compared to the enormity of the public education needs. These campaigns tend to be narrowly focused categorical efforts, often explicitly mandated and funded by Congress, on issues such as AIDS, breast cancer, high blood pressure, or substance abuse. To our credit, the federal government does target public education campaigns toward those at highest risk, among whom the "medically and informationally underserved" are disproportionately represented. Nevertheless, if we are to extend the benefits of public education to all those in greatest need, we will have to develop less costly and more effective mechanisms.

The activities and problems described above suggest certain core concerns in areas related to the accessibility, content, accuracy, reliability, and timeliness of health information. As the Public Health Service addresses the opportunities presented by new media, these concerns should shape its response.

THE FEDERAL ROLE IN SHAPING
THE NEW MEDIA ENVIRONMENT

The federal role in supporting the development of new media for health communication will be tied to two current federal initiatives, one to develop the National Information Infrastructure (NII) and the other to reinvent government through the National Performance Review (NPR). The intent of the NII initiative is to provide vision and direction for public and private activity in developing the national information superhighway.

Principles of the National Information Infrastructure

Several principles underlie the NII (Information Infrastructure Task Force, 1993):

- Promotion of *private sector investment* to develop both the infrastructure, and the applications to run on that infrastructure.
- Extension of the *universal service* doctrine from the regulation of access to simple telephone service to the regulation of access to information resources created as part of the NII.
- A role for the federal government as a catalyst for development of technologies supportive of NII deployment.
- Promotion of seamless, interactive, user-driven operation of the NII. The government is acknowledging its role as a guarantor of public access to a truly interactive "network of networks."
- Assurance of a secure and reliable information infrastructure which protects the privacy of its users and the integrity of the system.
- Improve the management of radio spectrum to assure that the wireless part of the NII is appropriately used.
- Protect the concept of intellectual property rights as more and more information on the NII comes as a result of individual or corporate effort.
- Integrate and articulate the NII with other local, state and international information infrastructures.
- The federal government will become a leader in the use of the NII for

the equitable and accessible dissemination of information to the public and in the procurement of technologies that promote the healthy development of the NII.

The NII outlines an impressive national agenda in the area of information and communication technology development and use. Although stating very clearly that the private sector will be responsible for financing most of the NII, the essential role of government agencies is made clear: to establish operation rules that ensure competition, access, and interoperability. Although receiving less attention from the health care community than the health reform debate, the implications of this initiative for the issues and applications discussed in this book are profound. As the "network of networks" envisioned in the NII develops, new relationships will form between new media applications for health and those supporting other domains of effort such as education, libraries, and government services.

Indications of how changes in government services will influence health-related communication technologies are found in the report of the national performance review initiative, the result of almost a year's effort at looking at how to apply contemporary customer-oriented management practices to a too-often cumbersome federal bureaucracy (Gore, 1993). Specific recommendations are included for each federal agency, including ones to the Department of Health and Human Services that encourage the development of integrated services and better use of contemporary information technologies. Integrated service delivery is at the heart of many new telecommunications-based, collaborative health applications. And, as described in several sections of this book, other technologies promise to enhance the ability of any agency, public or private, to provide better, faster and more comprehensive services to its customers.

THE PHS ROLE IN THE WORLD OF NEW MEDIA

The role of the Public Health Service in these two initiatives is to work with its federal partners, state and local public health agencies and the private sector to assure that sound principles of public health practice are incorporated into federal policies, activities, and investments.

The tantalizing vision of a national information infrastructure and a reinvented government holds great promise for reaching the public with health communication products and services. If we achieve the stated goal of wiring schools, libraries, clinics, and hospitals by the year 2000, this connectivity will permit a seamless, interactive, information-based health support system. Among the most promising elements of this new media system are universal (electronic) access to public health information resources and telecommunication-based outreach. In addition, interactive health education materials and decision support programs can be delivered either on-line or through stand-alone technology (PCs, kiosks, etc.). These approaches will be especially well suited to the future health landscape, where individuals will be increasingly involved in their own well-being.

A Framework for Federal Action

In this new media world, the array of available information-based health supports will have expanded, and the number of information providers will have increased. What then will be the role of the federal government? A framework for identifying priorities emerges clearly from the issues discussed above. The PHS and other agencies can frame their health information policies and activities around the core issues already mentioned: equitable access, especially for the traditionally underserved; core content; and selective quality assurance. *Healthy People 2000* lays out the opportunities for maximizing the impact of these information efforts.

Access.   Just as assuring universal access to health care is an accepted national objective, health information and education must also be widely and fairly available. In fact, given the often poorer health status of historically disadvantaged populations, the federal government has a special interest in ensuring that high-risk groups receive the benefits of sound health communication that is both passively assured and actively provided.

The administration's NII policy gives the federal government responsibility for the equitable and accessible dissemination of information to the public. The implication for the issues and applications discussed in this book are profound. As the "network of networks" envisioned in the NII develops, it is probable that a substantial amount of the health support enterprise will take place on this network. From interactive multimedia learning tools that enable pregnant women to better care for themselves, to telecommunications-supported home care that serves as an adjunct to—or even substitute for—nursing home care, the digital world of telecommunications connectivity gives new meaning to the term *cradle to grave coverage.* The federal government's role will be to ensure that there are no *have nots* in this new media support system. But access means more than just connectivity. It also relates to the content that is accessed or transmitted, and the ability of individuals to actually use the network in a comfortable, easy, and timely fashion.

Core Content.   If it is neither feasible nor desirable that the PHS attempt to provide all information to all people at all times, is there any collection of knowledge—including navigational tools like directories—for which the

PHS might assume an obligation to assure for everyone? Such a collection of priority content or core content might help address two problems: information overload on the part of the public, and financial constraints on PHS information providers. It should be an audience-driven approach based not on censorship but the identification of who needs what and when.

One can consider the issue of content at different value-added levels. Closest to the research source, the public health service will always have an obligation to disseminate findings from research supported with public funds. There is no clear mandate to carry the knowledge transfer process through to the ultimate end user in every case. Some guidelines do exist. The national prevention priorities are laid out in *Healthy People 2000*. Other special issues are identified by the president, the secretary of health and human services (and other agencies, such as the Environmental Protection Agency, the Department of Veterans Affairs) and congress. The PHS will always have, or share, lead responsibility for ensuring adequate production of health information and education content down to the appropriate end user in these areas. However, even in these instances, it may find valuable partnerships with other entities who can contribute special resources.

The private and voluntary sectors already play an important role in the health communication enterprise, especially in repackaging and delivering content derived from both Federal and other sources. Numerous private health information and education providers—some attached to health care providers—are already available to offer access to an infinite variety of health communication products and services. It is in the national interest to have as rich and diverse a set of information sources available to the public as possible. But, in the interests of a more rational health communication system, could we better delineate the most appropriate roles of all players? Is there an irreducible minimum content that the federal government has an obligation to assure for all Americans? Does the government have to actually provide this content directly to the public? Whether or not it is determined that this is in the public interest, are there ways of making all PHS information resources available more efficiently to other health communication intermediaries?

The private sector has always played an important role in the development and dissemination of health education materials. The creation of new media applications, such as those developed by ABC NewsInteractive, is costly and complex. It is likely that the private sector will be the largest developer of interactive multimedia health education programs in the future. However, the federal government will continue to provide leadership in several areas. First, it will still fund innovative model applications and their evaluation. Second, it will probably need to be a developer of new media materials in areas considered of national priority but not commercially attractive. Because the health risk behaviors addressed in *Healthy*

People 2000 include those identified as actual causes of death above, they are worthy of priority attention in the development of health-oriented communication technology. Some behaviors may be adequately targeted by communications applications developed through private sector investment based upon their likelihood of marketability. For example, interactive multimedia programs for smoking cessation or exercise counseling may find popularity as products purchased and offered by managed health care organizations for their enrollees, or offered on commercial telecommunications services into the home, school, or workplace. On the other hand, other behavioral risk areas might not be so readily addressed by private market forces. Violence or family planning issues, for example, may very likely continue to need public investment to attain optimal strategies of health risk communication. As new media technologies become more popular as a means of gaining the information necessary to change behaviors, PHS investment in applications that address these "orphan risk behaviors" will be essential.

Accuracy, Reliability, and Timeliness.   Accuracy, reliability and timeliness are clearly prime determinants of the value of all health communication products and services. The federal government is responsible for ensuring the integrity of its own information. However, all government information is in the public domain. We cannot control derivatives, only impose minimal requirements on how federally-produced information is used. In a new media environment, the federal government will still be responsible for quality assurance for its own content. The government already insists that repackagers keep its material intact, identify it as having been developed by the federal government, and (often) include a statement that the federal government does not endorse any particular product or service. These guidelines can be adapted to ensure the integrity of government-generated information in electronic applications developed by others.

Additionally, it may be that the federal government adapts its current practice of providing content review and assistance to the private sector, validating the work of commercial applications developers without specifically endorsing their product. Any efforts to identify appropriate roles for the federal government and other health communication providers will need to address this question of quality assurance all along the chain of information production and communication.

The Public Health Service, like other federal agencies, faces resource constraints and internal procedures that limit our ability to deliver information regularly and on time. These limitations are the greatest when there is a significant value added component, as with public education materials, or policy implications, as with medical practice guidelines. Others' repackaging and dissemination of our content (as well as their own) is often invaluable in speeding delivery to all those who need it quickly. In the future, we should explore partnerships with various electronic information developers and/or providers, including digital publishers and online services, as well as direct internet dissemination.

CONCLUSIONS

To play its role fully in the new environment, the federal government will have to examine its own procedures and assumptions about its health information activities. It will have to decide on the value of this information and devise structures that support its health information goals. Individual agencies will have to develop less proprietary attitudes toward their own health information and consider approaches combining their special expertise with other resources for both content and delivery.

To help ensure that the benefits of the new media for health information are widely and equitably extended, the federal government may have to assume other responsibilities not yet anticipated or fully understood. For example, basic technology literacy will be vital if the new media are to be used to their fullest—and most appropriate—potential.

But to accomplish any of this, the federal government must first educate itself about the new media and health. It must understand the rich array of information-based interventions that support our national health goals. It must understand which media support which interventions. It must understand who needs to know about these new media health applications and how to educate them. Not surprisingly, these new health communication media begin and end with education.

REFERENCES

Ashton, J. (1992). Handling information as a professional. In A. van Berlo & Y. Kiwitz (Eds.), Information in a healthy society: Health in the information society (pp. 17–21). Knegsel, The Netherlands: Akontes.
Clinton, W. (1993, February 22). Speech unveiling administration's new technology plan. Washington, DC.
Deering, M. J. (1994, January). [Preliminary survey of public health service clearinghouses]. Unpublished data collected for planning meeting, of conference on networked health information for the public.
Farquhar, J. W., Fortman, S. P., Flora, J. A., Taylor, C. B., Haskell, P. T., Williams, P. T., Maccaby, N., & Wood, P. D. (1990). Effects of communitywide education on cardiovascular disease risk factors. Journal of the American Medical Association, 264(3), 359–365.
Fecher, E. (1985, February). Consumer health information: A prognosis. Wilson Library Bulletin, pp. 389–391.

Gore, A. (1993). *From red tape to results: Creating a government that works better and costs less.* Report of the National Performance Review (Report No. 040-000-00592-7). Washington, DC: U.S. Government Printing Office.

Gore, A. (1994, January). *National telecommunications reform.* Speech to Academy of Television Arts and Sciences, University of California, Los Angeles.

Harris, J. (1994, January). *Preliminary study: Consumer demand for health information.* Unpublished manuscript prepared by Reference Point foundation for planning meeting of conference on networked health information for the public.

Information Infrastructure Task Force (1993, September). *The national information infrastructure: Agenda for action.* Washington, DC: The White House.

McGinnis, J. M. (1988–1989, Winter). National priorities in disease prevention. *Issues in Science and Technology, 5*(2), 46–52.

McGinnis, J. M., & Foege, W. H. (1993, November). Actual causes of death in the United States. *Journal of the American Medical Association, 270*(18), 2207–2212.

Mullan, F. (1989). *Plagues and politics: The story of the United States Public Health Service.* New York: Basic Books.

National Capitol Systems, Inc. (1986, January). *Final report on assessment of users' satisfaction with seven PHS clearinghouses.* Prepared for Department of Health and Human Services, Public Health Service, Office of Public Affairs.

Rees, A. M. (1991). Medical consumerism, library roles and initiatives. In A. Rees (Ed.), *Managing consumer health information services* (pp. 23–36). Phoenix, AZ: Oryx.

U.S. Department of Health and Human Services, Centers for Disease Control and Prevention, National Center for Health Statistics (1993). Advance Report of Final Mortality Statistics, 1990. *Monthly Vital Statistics Report, 41* (Suppl. 7).

U.S. Department of Health and Human Services, Centers for Disease Control and Prevention, National Center for Health Statistics (1993). *Health United States 1992 and Healthy People 2000 Review.* Washington, DC: Government Printing Office.

U.S. Department of Health and Human Services, Centers for Disease Control and Prevention, National Center for Health Statistics (1993, April). *Years of potential life lost before age 65: United States, 1990 and 1991.* In *Morbidity and mortality weekly report, 42*(13), 251–253.

U.S. Department of Health and Human Services, Office of the Assistant Secretary for Public Affairs (1985, May). *An examination of information clearinghouses within the Department of Health and Human Services.* Unpublished manuscript, submitted to the Office of Management and Budget. (In Accordance with OMB Memorandum 81-14, 1/11/1981).

U.S. Department of Health and Human Services, Public Health Service (1991). *Healthy people 2000: National disease prevention and Health promotion objectives.* Washington, DC: Government Printing Office.

U.S. Department of Health and Human Services, Public Health Service (1993). *1992 National survey of worksite health promotion activities: Summary report.* Washington, DC, Government Printing Office.

U.S. Preventive Services Task Force (1989). Guide to Clinical Preventive Services: An assessment of the effectiveness of 169 interventions. Baltimore, MD: Williams & Wilkins.

Will, G. (1989, January 8). Cocaine: A trickle-down affliction. *Washington Post,* C7.

IV

HEALTH EDUCATION

Comprehensive School Health Education and Interactive Multimedia

Sandra Cheiten
ABC News InterActive

Mae Waters
Florida Department of Education

Many schools have been in the health care reform business for several years. Throughout the U.S., state departments of education and school boards are emphasizing comprehensive school health programs to provide health education and services. The goal is to keep students healthy, prevent those behaviors that are leading causes of morbidity and mortality, and promote behaviors that lead to healthy lives. At the federal level, comprehensive school health education programs have been integrated into the national education goals under Public Law 103–446 passed in May, 1994. Programs under this law are mandated to ensure that all students receive quality comprehensive school health education by the year 2000.

Providing for the health needs of students is a special challenge for health care reform. Appropriate instruction of our children gives us a perfect medium in which to stress prevention education. This effort will lower future health care costs by means of early diagnosis, treatment, and modification of health risk behaviors.

Health Risk Behaviors

The Centers for Disease Control and Prevention (CDC) is one of the federal agencies that works with both state departments of education and health agencies for funding and implementation of prevention education. The CDC has identified six of the most common health risk behaviors to target in adolescents in school health programs. It has provided support and funding to 10 state departments of education to establish a comprehensive school health program (CSHP) to integrate efforts to reduce risk behaviors in each of the following six categories: Behaviors that cause unintentional and intentional injuries; drug and alcohol use; sexual behaviors that encourage sexually transmitted diseases (STD's), including HIV infection, or lead to unintended pregnancies; tobacco use; inadequate physical activity; and dietary patterns that cause disease.

Each of these six behaviors should be addressed by eight identified components to ensure proper preventative health education. The following components should be addressed in each state's comprehensive school health program: health education; health services; biophysical and psychosocial environments; counseling, psychological and social services; integrated efforts of schools and communities to improve health; food service; physical education and physical activity; and health programs for faculty and staff.

Whereas the health problems of young children are often related to the conditions of their mothers and families, the leading causes of death among adolescents—injuries from motor vehicle crashes, suicide, and homicide—are mainly the result of adolescents' risk-taking behaviors. Other health problems such as sexually transmitted diseases, teenage pregnancies, substance abuse and injuries are often a result of risk behaviors, and are more common among adolescents than among younger children or adults. Similarly, many diseases and other illnesses experienced later in life are a result of behaviors begun during adolescence such as smoking, lack of exercise, and poor diet.

PREVENTION EDUCATION AND INTERACTIVE MULTIMEDIA

Many of these health risks are being addressed, using interactive multimedia health education. New interactive multimedia health education programs allow the teacher and students to not only get factual information, but to experience the material in a personal way.

What Does it Mean for Students?

It is very important to acknowledge that young people view many hours of television daily. We need to realize the importance of providing television quality multimedia to schools that not only is entertaining but are curriculum based. For today's children, who have been raised on television, traditional textbooks, chalkboards, and lectures can be boring. Interactive multimedia can motivate a student to respond and learn.

Through multimedia programming, students, regardless of learning styles, can access accurate information that is created in the style of the television shows they are accustomed to watching. This means the visuals are engaging and catch the attention of the students, accommodate short attention spans, and can be accessed repeatedly by the students in order to master the material before moving on. The ability to access visually compelling materials that are personally involving, enhances the learning process.

When students use multimedia individually, they have confidential access to health information. If they choose to learn about birth control, abortion, the correct way to use a condom, or how to deal with an alcoholic family member, they have entree to these materials without anyone controlling what they are learning. If they choose to view the material more than once to fully understand the content, they can. Students can acquire information at their own pace, without having to ask personal questions in front of teachers or their peers.

Adding computerized hypermedia makes even more sophisticated learning possible (as described in more detail in chapter 9). Hypermedia automatically cross-references archived information, enabling students to do independent research using the original source materials provided, as well as creating video reports which reference video segments. By using hypermedia students can access and organize support material, glossary terms, and background information to support the visual message from the videodiscs. Hypermedia tools make it possible for students to be producers of interactive instruction. Learning by teaching can stimulate student interest, increase comprehension, and improve critical thinking and organizational skills.

There are many possibilities to enhance cooperative learning (Carlson & Falk, 1989). Students working together on a project teach each other, learn to share, and experience teamwork. Each multimedia product contains numerous suggestions for these types of activities. Students can be trained to use multimedia for peer education. Young people can provide accurate information to their parents, peers, or to younger students. In health education, peer education has been heralded as an important tool for imparting information and providing positive role models.

When students produce their own multimedia presentations, they demonstrate their newly acquired knowledge. The accuracy of the material the students work with can be professional verified. This kind of production-based learning experience can involve the preparation of messages for their peers, younger students, or to take home to their parents, thereby extending learning from school into the home. This facilitates the extension of accurate health information into the home and the community. Students have used videodiscs to work in peer education situations, such as a classroom setting, or in individualized instruction, or in counseling situations.

The use of a *documentary maker* function (similar to cutting and pasting text) allows students or teachers to create their own video presentation from the information archive. There are several types of items that can be in-cluded in a documentary: video clips (video segments from the videodisc in their original length), custom clips (video segments that are created from the full video clips already on the videodisc), titles (text that can be sent from the computer to be displayed on the video monitor), notes (comments that can be used during the replay of the documentary or a report accompanying the presentation), sounds and pictures. A documentary can be exported for someone else to use. By accessing any chapter, or part of a chapter, still frame, chart, graph, or fact file, a lineup can be created and stored in the computer to be accessed like a moving video or played individually. A video can be created to answer a question or create a video report.

What Does this Technology Mean for Teachers?

Few technologies have made an impact on education as forcefully and as quickly as interactive multimedia (Fletcher, 1989, 1990). For many, the acceptance of multimedia as a teaching tool by health educators comes as no surprise to anyone who has seen the power of this technology in action. It is an instructional tool that allows teachers to take control of the learning environment in an entirely new way. It excites students to learning impor-tant information that will impact their behavior. An interactive multimedia documentary can be a video lesson plan that can be designed, used, refined, and reused.

In 1990, Texas made history by approving textbook funds for the pur-chase of videodisc hardware and software. Since the Texas adoption, text-book publishers have realized the necessity of including multimedia with their products. Some are producing multimedia to accompany their text-books, realizing the power of multimedia to enhance learning and provide the teacher with a video database to reinforce their lessons. Many new textbooks are being printed with barcodes in their margins and reference videodiscs and other multimedia.

According to Market Data Retrieval, a Connecticut-based firm, the number of computers in the nation's elementary and secondary schools reached 2.5 million in the 1991–92 school year (personal communication, 1990). Thus, an ever-increasing number of students and adults are being exposed to quality health education through multimedia.

Multimedia have played a major role in restructuring how educators teach and how students learn about health. Multimedia health education fosters and supports active learning, cooperative learning, individualized learning, and interdisciplinary learning. Teachers using multimedia have reported that students become active learners and participants, rather than remain passive learners.

The random access format allows educators to choose the order and pace at which information is presented. This technology encourages teachers to customize the video material to the academic level, community values, curriculum needs, and interests of their students. Teachers can pause at any time to answer pupil questions, reinforce a point, analyze a graphic, and stimulate discussion. Although today there are few results to validate the effectiveness of interactive technologies in the health education domain, a number of research studies are being conducted by various educators and universities. Studies in various other education domains have demonstrated the effectiveness of interactive technologies. (Telecommunications Development Center, 1989).

In the past, videos and movies were often shown in dark rooms with both the teacher and student inactive. The use of this new technology allows the teacher to manage the media. It puts control of the media into the hands of the teacher (Carlson & Falk, 1991). The use of interactive multimedia has provided a new direction for many health educators to disseminate important, medically accurate information to young people dealing with critical health issues in a new and exciting manner. Today the teacher and students can control the learning environment through selecting what they want to view, when they want to view it (Moore & Carnine, 1989; Slike, 1989).

Individualized instruction can be created by using the lesson plans provided for the teacher or students, thus the needs and skills of individual students can be addressed. The teacher can create new types of evaluation measurements by allowing the students to submit video reports that have been created using video material from the videodiscs. Students' progress on their learning from visual sources can be assessed and supported by incorporating resources from the software. Students may review material many times to discern what is most appropriate for their report. It is a new way of learning and assessing student progress in nontraditional ways. Student assessment can be creative, by assigning students video reports, supported by a written report created from sources in the software. Tests can be initiated by asking questions after viewing video chapters, and teachers can have students answer the questions provided in the software as an assessment tool. Educators can also create their own barcodes by computer, enabling them to select and sequence material to accompany their own printed material, text, or personally designed lesson plans.

With hypermedia and/or networking software the user has access to legislation, articles and documents, all of which may be printed from the student's desktop. Users become "master presenters" of information. They have the ability to include new material, which allows for the sources of information to be continuously updated in the rapidly changing health field. These multimedia products are a portable interactive source of health expertise. They can be taken and shared with any audience or group.

The use of interactive multimedia health education programs has enabled teachers, who are required at times to teach sensitive material such as AIDS, sexuality and drugs to use multimedia to present needed health information. It can be difficult for some teachers to discuss these subjects. In such cases teachers can bring compelling information to students delivered by persons such as the surgeon general and medical professionals. Personal stories of young people are an engaging way to impart knowledge. When students hear other students talk about personal experiences dealing with topics such as sex, drugs, tobacco, alcohol, it encourages them to make healthy decisions.

Multimedia programs can add value to school health education in the following ways. As *lecture aids*, videodiscs provide visual support for lesson plans. They enliven lessons and encourage classroom discussion. Using a *comprehensive visual archive*, students and teachers have immediate access to compelling information stored as visual images. Students can incorporate video from these videodiscs into their written reports or oral presentations. Multimedia programs are also a *vehicle for application of student computer skills*. Students learn computer skills by developing their own research projects, reports, and other interactive applications. It allows students to develop individual presentations that can be shared with their peers and parents. They can gain through *individualized learning*. Students who desire enrichment or independent study can spend time reviewing the lesson materials at their own pace. Multimedia is also a *vehicle for peer teaching*. This cutting edge technology excites and motivates students. It encourages them to take an active role in their own learning process and to share their enthusiasm with fellow students. It becomes a *platform for cooperative learning*. Students can work in groups to develop their own videos and to discuss health information and behavioral strategies with their peers.

The value of multimedia use in schools has spearheaded the implementation of prevention education throughout the curriculum. Many educators feel that it will have a tremendous impact in health status, by allowing students to improve their risk behaviors. This in turn could have a direct effect on cost savings for the medical community. If the risk behaviors of young people are addressed, health problems can be avoided, not only effecting medical costs but saving the lives of today's youth.

ABC NEWS INTERACTIVE

A wide spectrum of people, drawn from many professions, disciplines, businesses and organizations are grappling with the opportunities and challenges of interactive multimedia. One of these organizations, ABC NEWS, began to tackle this challenge in 1988 when it created *ABC NEWS InterActive*. The original purpose of the unit was to recycle ABC NEWS archival materials, and make them available for use in schools. The objectives for the unit were soon broadened and today, *ABC NEWS InterActive*, is dedicated to producing multimedia in both health education and social studies. ABC NEWS archives serve as the basis of the programming with enhancements from original productions. There is additional support by research in conjunction with agencies such as the Centers for Disease Control and Prevention, national health organizations health officials, and health educators on the national, state, and local level.

ABC NEWS InterActive's multimedia are designed to support and enrich classroom instruction. Combining news, data, and personal stories, the random-access format of multimedia allows educators and health professionals to present information in any order and at any pace they choose. Current research is underway to verify the effectiveness of multimedia in improving student learning and changing student behaviors. *ABC NEWS InterActive* provides teaching tools to help educators and health information providers present complex topics and ideas in a new and exciting manner.

After its first year of producing social studies, *ABC NEWS InterActive* began a health series, *Understanding Ourselves*. *AIDS* was the first title. Dr. C. Everett Koop agreed to be the medical expert and on-screen spokesperson and has worked with *ABC NEWS InterActive* on four titles. *ABC NEWS InterActive* has now produced a series of six titles in health education: *AIDS, Drugs and Substance Abuse, Teenage Sexuality, Tobacco, Alcohol ,* and *Food and Nutrition.* Two new health titles are currently in production which will complete the subjects in comprehensive health education to include the six health behaviors as defined by the Centers for Disease Control and Prevention.

Accompanying *ABC NEWS InterActive* multimedia, are curriculum-based support materials that include lesson plans and cross curriculum integration charts that extend the use of the multimedia in and out of the classroom. *ABC NEWS InterActive* provides students in health education with contextual, real-world application of materials. It allows students to gain hands-on experience with subjects some teachers may find too troubling or are uncomfortable with. The multimedia content can also be used in prevention health education programs by local health providers in health centers and in hospitals, thus extending the potential users of these products.

ABC News InterActive's multimedia health education series ensures a standard quality of accuracy for the information delivered. Within a state or school district, health educators have the ability to standardize the delivery of materials for specific grade levels integrating curriculum objectives with community standards. This means the materials deemed educationally appropriate to the age level and content standards of the community can be maintained. With barcoded lesson plans developed at the local level, a community can ensure a uniformity of lessons. In one school district, where it was forbidden to teach about the use of condoms, the introduction of the *AIDS* videodisc, with Dr. Koop discussing the use of condoms, broadened the community acceptance of this topic and eventually led to the inclusion of this topic into the local curriculum. Multimedia allow local educators to display accurate information in conformity with local needs. The health series, *Understanding Ourselves*, has a multiplicity of uses. It can be used not only by grades K-12, but can be utilized by PTA and community groups.

The *ABC NEWS InterActive* multimedia health education program currently consists of a two-sided laser videodisc and supportive print material that includes lesson plans and hypercard stack. The two-sided videodisc contains a half-hour of moving video on each side and hundreds of fact files, maps, graphs, charts, glossary terms, and additional information. The videodiscs can be viewed with English or Spanish audio tracks and are closed-captioned for the hearing-impaired. The videodiscs are accompanied by a software program available on Macintosh and MS–DOS platforms. The software contains lesson plans, statistics, interviews, primary text documents including speeches, biographies, and reports. It also allows the user to cut and paste from the videodisc to create their own video. This tool can be used by the teacher to develop lesson plans and by students to design their own video. The software is in Spanish as well as English. A comprehensive set-up and operational manual is also included. A printed guidebook is also included, containing a barcode compatible directory of the videodisc's contents and lesson plans, cross curriculum index, and enrichment questions and activities.

The components necessary for using *ABC NEWS InterActive* multimedia include a videodisc, a videodisc player, and a television monitor. Using the remote control a teacher is able to establish the pace of any lesson and access any chapter or frame of the videodisc. The information on the videodisc can also be accessed with the use of a barcode reader. Adding a computer to this system creates many more applications.

Although *ABC NEWS InterActive* is currently producing a series in health education on the videodisc platform, they are looking to other formats such as CD-ROM and networking to deliver its material and extend the health education series and provide broader access to important health information. *ABC NEWS InterActive* sees itself as content providers, regardless of the platforms utilized.

THE FLORIDA MODEL: PARTNERING WITH *ABC NEWS INTERACTIVE*

The use of multimedia in health education was pioneered in Florida. Florida ranks among the top five U.S. states with mortality rates greater than 1,000 deaths per 100,000 population (Centers for Disease Control and Prevention, National Center for Health Statistics, 1989). In 1990, Florida had 10,100 deaths per 100,000 population (Florida Vital Statistics, 1990). The leading causes of death for all ages and ethnicities combined in Florida are: heart disease, cancer, stroke, chronic obstructive lung disease, and accidents. HIV was the 4th ranking cause of death for nonwhites whereas it ranked 10th for whites (for the combined population it ranked 8th).

The Florida Youth Risk Behavior Survey, administered to a sample of students in grades 9 to 12 in April 1991 revealed that high percentages of those students were engaging in health risk behaviors (Waters, 1991). Tobacco was the second most used substance reported by the sampled students following the use of alcohol. Seventy-two percent of all students in this survey have tried smoking cigarettes. More females (29%) than males (25%) reported they have smoked regularly (at least one cigarette a day for 30 days). The percentage of students who have smoked regularly increased from 9th to 12th grades (25% in 9th grade; 31% in 12th grade).

One of the main organizations to focus on the problems of tobacco is the Florida division of the cancer society. Through their codevelopment efforts with *ABC NEWS InterActive*, a multimedia product on tobacco was created. Teachers have indicated that by integrating *Tobacco* into their instruction, students have been able to focus on their behaviors. Nontobacco using students become positive role models. Through the blend of news stories, personal vignettes, and information supplied by the cancer society, teachers report that *Tobacco* allows students to evaluate the power of the tobacco industry and the power of tobacco advertisements to entice young people to use tobacco products. Through multimedia, students are able to analyze the media through the media, therefore becoming increasingly media literate.

Many of the students in the Florida survey did not participate in physical activity nor did they indicate that they ate healthy diets. Fifty percent of all sampled students were enrolled in physical education classes. Males participated more often in vigorous physical activity than did females (78% for males and 54% for females). The amount of physical activity declined for both sexes as they progressed in grade levels. Physical inactivity is a major risk factor for cardiovascular diseases and promotes premature mortality. Children and adolescents can reduce their risk for later heart disease through regular physical activity. A physical activity multimedia project is now being considered for future production. This multimedia product would complete technology instruction related to the six risk behaviors identified by the CDC.

Florida students often ate fatty food and sweets such as doughnuts, cookies, and cake rather than vegetables and salads. Sixty percent reported eating French fries or potato chips the day before the survey and 57% ate sweets. Less than half (46%) of the students ate cooked vegetables and less than a third (30%) ate green salad the day prior to the survey. Instructional strategies developed and found in the curriculum related to the *Food and*

Nutrition program address these and other issues. *Food and Nutrition* also explains the new food label and the food pyramid which allows teachers to use technology to reinforce and update textbook instruction. Through exciting graphics, current nutritional information is presented in ways to inform individuals about making healthy choices at fast food restaurants, in selecting groceries, and in making healthy eating choices. Dr. Marcia Angel, of the *New England Journal of Medicine*, provides medically accurate information that is accompanied by ABC NEWS archival material and original production of interviews with student athletes and health professionals from across the U.S. on various related topics.

Sixty-one percent of all students in the Florida survey reported that they had engaged in sexual intercourse (79% for 12th graders). Over half (54%) of students surveyed who had been sexually active did not use a condom the last time they had sexual intercourse. Thirteen percent of the 12th-graders in this survey who were sexually active reported having been pregnant or had gotten someone pregnant. This study also found that more 9th-graders in the Florida sample started smoking regularly, drinking alcohol, and engaging in sexual intercourse at a younger age than did students in grades 10 through 12. Teachers have found that by using *Teenage Sexuality* as an integral part of classroom instruction, student involvement in peer instruction has allowed for improved communication between students and teachers. This particular tool was the leading instructional strategy in Florida's implementation of K-12 mandated sexuality legislation.

Adolescents aged 15 to 19 have higher rates of syphilis and gonorrhea than the rates for the population as a whole. Instruction from medical settings on *Teenage Sexuality* allows the classroom teacher to use experts to reinforce classroom learning. The surgeon general speaks directly to the class. Medical procedures and counseling efforts are demonstrated. Multimedia allows for individual involvement in prevention education and addresses some of the issues highlighted in the current discussions of sex education as part of health care reform.

Many parents became aware of the need for HIV-AIDS education when they participated in a review session to provide parental support for classroom instruction. *AIDS and Teenage Sexuality* provided medically accurate information from physicians and the surgeon general and are used to influence young people on the dangers of risky behaviors.

The consequences of these risk behaviors are teenage pregnancies, sexually transmitted diseases including HIV, and babies born with HIV or long-term health problems. Florida has the second highest rate of AIDS among the states. Florida's rate of AIDS cases for the year ending March 1992 was 42.1 cases per 100,000 (U.S. average rate is 18.1). Twenty-one percent of the AIDS cases were for the age category of 20 to 29 years old. Because it can take up to 10 years for AIDS to develop, infection with HIV probably occurred when these young adults were teenagers. As of 1990 the leading cause of death among young black females age 15–44 is AIDS. For males aged 15–44, AIDS was the second leading cause of death. Health educators were among the first to bring technology into the classroom by using the AIDS multimedia product. Dr. Joycelyn Elders, former U.S. Surgeon General has said that education is the only vaccine we have against the AIDS virus.

Florida Teachers and *ABC News InterActive*

In January 1990 at the Educational Technology Conference in Daytona Beach, Florida, members of the Florida Department of Education Prevention Center were introduced to *ABC News InterActive*'s *AIDS* multimedia. This sparked their interest in the use of multimedia for the teaching of AIDS and human sexuality. *ABC News InterActive* was contacted and scheduled to demonstrate the technology for members in comprehensive health education in Florida.

The *ABC NEWS InterActive AIDS* multimedia team arrived in Tallahassee, Florida, to showcase *AIDS* to the Department of Education's AIDS Advisory Council and a small group of district comprehensive health education coordinators. Those present for the demonstration were highly impressed and enthusiastic concerning the prospective advantages for Florida's teachers and students. Thus the relationship between *ABC NEWS InterActive* and the state of Florida began as an effort to bring multimedia into health education classrooms.

ABC News InterActive, Apple Computers, and Pioneer Communications showcased the technology to Florida's comprehensive health education/drug-free schools coordinators at the February 1990 grant writing workshop in Tallahassee. Participants were asked to indicate their level of interest in participating in an AIDS multimedia pilot project, should one be developed. The response was overwhelming. Coordinators submitted written requests to participate in a multimedia project before they even left the demonstration.

The AIDS interactive education project was developed, and detailed implementation arrangements were negotiated. *ABC News InterActive* supplied *AIDS* to pilot projects and assisted in training. Apple Computer supplied Macintosh computers for training sessions and arranged a loan of those computers to the pilot projects for the duration of the project. Pioneer Communications supplied Pioneer videodisc players for training sessions and also arranged a loan of the videodisc equipment to pilot projects.

The AIDS Project secured funding from the CDC and developed the training sessions for the implementation of this exciting new technology. Funding was sufficient to establish an AIDS multimedia pilot project in each school district that had indicated an interest. The pilot projects encompassed 16 Florida school districts, two education consortia (representing 15 school districts), three model technology high schools, one community college, and one university.

Training for the pilot projects' comprehensive health education/AIDS district coordinators, model technology high schools' representatives, and university representatives was scheduled to go to Tampa and the training agenda and teaching lesson plans were developed. Specialized evaluation instruments were also developed for the state-level training session and for evaluation at the pilot project level.

Training for the 23 participants was accomplished in two stages.

Stage One.   The first session, held in Tampa, May 21–22, 1990, trained participants to use the hand-held remote control to access any segment of the videodisc and to learn to use it as an educational tool. Participants also learned to use a barcode reader as a quick way to access segments of the videodisc. The second, held in Sarasota, July 16–18, 1990, trained participants to use the software created by *ABC NEWS InterActive.*

It was expected that by the end of the two training sessions, the participants would be able to operate multimedia well enough to train others in their institutions. This expected expertise also included being able to utilize sample lessons devised by the department of education and *ABC NEWS InterActive* staff and to design their own lessons using multimedia. The trainees were given evaluation forms developed by the Florida Department of Education and were frequently monitored by DOE prevention department staff. All reported success in their training and in training fellow health educators in their districts.

Over time, the training has been modified and redesigned to better meet the learning needs of educators. Much of the training methodology has improved. It is now possible to train teachers to use Level 1 videodiscs within an hour, and to have them developing their own lesson plans in a half-day session. The use of Level 3 training depends on the expertise of the teachers on computers. If teachers are computer literate, it takes half a day to a full day for teachers to master the software and be able to create their own lesson plans and develop their own documentaries. If the trainee has no computer training, it takes three days of training, much of it devoted to introduction of the computer.

The training is now conducted in a Training of Trainers Model and is provided at the state level by a program specialist within the department of education's comprehensive health program. Current training provided by the state is offered on a half-day or full-day basis for Level 1 interactive laser discs use. Half-day instruction allows for introducing the technology, mastering basic operational skills, and a short previewing session for participants. Full-day instruction allows for more in-depth practice of skills, a more focused previewing session to allow participant time to familiarize themselves with the disc and its operations and time for participants to develop actual classroom lessons for use with students. The full-day instruction is optimal, as participants in these sessions are more likely to feel confident that they can use the technology in their classroom when they leave the training session.

Computer training sessions have also been modified and streamlined. A full day of training will allow the basics to be accomplished with first-time computer users. However, additional practice time on the part of participants is necessary for them to feel at ease with the technology. This additional time is most effective if the time is scheduled for participants to formally meet for a practice lesson plan development session after participants have had individual time to practice.

"The first time I showed the videodisc on AIDS was a Wednesday morning in November," recalls Dianne Douglas, health teacher at John I. Leonard High School in West Palm Beach, Florida. Douglas remembers that about halfway through the disc, a profile of Amy Sloan, a midwestern housewife with AIDS, was just concluding with Amy's own words: "I didn't do anything to get AIDS." "There was total silence in the room," remembers Douglas. "Nobody moved, and nobody spoke. I didn't say anything. I let the silence sink in. The students sat there just spellbound."

Stage Two. The second phase of this project began with Former Commissioner Betty Castor providing a videodisc player for each school in the state. The availability of the videodisc players prompted numerous requests for training in the utilization of this new technology. To assist with the growing demand for training, the department of education awarded the grant to Broward County Schools Health Education section, one of the original interactive education pilot project sites, for the development of a step-by-step training manual and video for the computer portion of the program. The manual is used in training sessions as a tool to assist participants in making their way through the development of a computerized lesson plan. This manual and videotape are used as guidebooks for participants to take back to their schools and classrooms as reference documents.

Training requested from the Florida Department of Education's Comprehensive Health/AIDS Education staff takes the form of both training for trainers and training for teachers. The training includes hands-on experience both with technology and content. The training for trainers sessions provided by this staff are designed to provide participants with the knowledge and expertise needed to conduct sophisticated HIV/AIDS education and human sexuality teacher training utilizing multimedia. The training for teachers sessions prepare classroom teachers for using this technology with their students. Upon completion, classroom implementation of HIV/AIDS education and teenage sexuality multimedia lessons is within the grasp of training participants. The technology seems to have sparked the enthusiasm of the teachers to provide much needed content information on these essential health topics in a new and exciting manner. Teachers are leaving the training sessions enthusiastic, to return to their districts and classrooms and begin using the information they have received to augment their existing teaching materials and styles.

Beyond bringing innovative HIV and human sexuality education to students, these projects have broken ground for other health education topics to be taught via this highly interactive medium. Teachers look forward to the next multimedia product to make a dramatic impact on their teaching and therefore on the health and lives of students in Florida and throughout the nation.

Interactive multimedia was designed especially for classroom teachers as a means of enhancing and ultimately revolutionizing classroom learning. Multimedia products have been termed *living textbooks*, and are unique in that the classroom teacher is able to randomly access a vast array of video footage divided into chapters, charts, maps, and fact files which are appropriate for the students in that classroom. A classroom teacher has the ability to present only the materials deemed appropriate for his or her students based on their age, maturity, and school board and community guidelines. By virtue of the selection of specific frames and chapters, this is accomplished. If some information in the product is too difficult or inappropriate for a particular class, it need never be accessed. If some information in a chapter needs to be repeated it can be accessed repeatedly. Thus the flexibility of multimedia for health education allows them to be utilized in a range of environments and age groups. "It has stimulated my teaching style. It has been a jump start to new directions. I am respected in the eyes of my colleagues in other subjects," stated Carol Tasca, Health Teacher, Christiana High School, Delaware.

This innovative project has made Florida a national leader in utilizing multimedia to assist in health education instruction. The multiorganizational team including the department of education's comprehensive Health/AIDS Education Project staff and teams from *ABC News InterActive*, Apple Computers, and Pioneer Communications of America, developed a training program that will serve as a model for other states and health education programs throughout the nation. Since the original training, other states such as Texas, Rhode Island, Delaware, Kentucky and California have had statewide training of health educators from their regional centers in the use of this multimedia technology. "It has been one of the best received projects that I have promoted for teachers and administrators throughout the state of Delaware," said Edith P. Vincent, Education Associate, Department of Public Instruction, State of Delaware.

Comprehensive school health education programs have been integrated into the national education goals as identified in Public Law 103–446, passed in May, 1994. Programs under this law should ensure that all students receive quality comprehensive school health education by the year 2000.

NATIONAL INVOLVEMENT

The *ABC News InterActive* series *Understanding Ourselves* was created with a national advisory panel to ensure its quality and medical accuracy. Nationally recognized experts such as Dr. C. Everett Koop, Dr. Joycelyn Elders, and Dr. Marcia Angel, are called upon to lend their expertise in specific health education areas. Medical associations, research institutions and federal agencies such as the CDC, National Institute of Health, U.S. Department of Education, U. S. Department of Agriculture and many more are used as definitive sources for the latest medical information and statistical updates. The guidebooks are reviewed by leading health educators from states such as Florida, California, New York, and Texas.

There has been great excitement by national health education organizations to promote the use of multimedia technology in the teaching of health education. Many national organizations have highlighted the *ABC NEWS InterActive* health series to showcase health educators as leaders in the use of technology in education. The American Association of Health Education (AAHE), supported by funds from the Metropolitan Life Insurance Company, launched a national awareness program to enlighten health educators in regard to the latest trends in health education. Four exemplary health models were discussed at six national AAHE meetings in the Spring of 1991, They were attended by local, state, and higher education health educators from 50 states. *ABC NEWS InterActive* was selected to showcase "*Understanding Ourselves*" and the use of multimedia in health education. Through these national meetings, health educators throughout the nation were exposed to the use of interactive technology to teach health education. From the evaluation forms returned at the end of the meetings, to the letters and comments received by the American Association for the Advancement of Health Education, the response was tremendous. Health educators became aware of this new technology and its power to enhance and enliven classroom and individualized instruction.

According to Dr. Rebecca Smith, Executive Director of the American Association for the Advancement of Health Education, "When funding was made available to the Association for the Advancement of Health Education by the Metropolitan Life Foundation to hold workshops across the nation to strengthen health education instruction, the *ABC NEWS InterActive* computer and laser discs technology was one of four cutting-edge instructional techniques presented. We strongly believe that health education instruction will benefit from more creative use of new technologies." (personal communication, 1991)

Through presentations at various state and national health education conferences, teachers, local and state level health educators, and other health professionals have become aware of the use of interactive multimedia for health education. Adoptions of the *ABC NEWS InterActive* health series have occurred in Rhode Island, Texas, Delaware, California, and Texas. Multimedia, hardware, and teacher training have been provided by AIDS categorical grants from the CDC funds and by Drug Free School grants from the U.S. Department of Education.

Through a codevelopment effort of *ABC NEWS InterActive* and the Florida chapter of the American Cancer Society, a multimedia project, *Tobacco,* was developed. This project was the result of a partnership between these organizations to create a visually exciting and factually accurate multimedia tool. A national advisory panel of experts was jointly established to provide *ABC NEWS InterActive* with the most current and relevant material on tobacco.

Dr. C. Everett Koop, once again, was the spokesperson and Ted Koppel provided the overviews. ABC NEWS archival footage was utilized, as ABC NEWS has been covering issues dealing with smoking and tobacco use for years. Included was the original press conference in which the surgeon general announced warning labels would be placed on all cigarette packages to television commercials for Philip Morris and the original Marlboro Man. Special segments were created with ABC NEWS correspondents. Original shootings were held in Florida, Texas, and California, where students were interviewed discussing their attitudes toward smoking, smokeless tobacco, self-esteem and ways to create a smoke-free environment. Not only does *Tobacco* include information on the short- and long-term effects of tobacco, including smokeless tobacco, it also provides a puppet show for young students shot in California, and a rap promoting a smoke-free lifestyle. Student interviews as role models are used to stimulate classroom discussions hopefully leading to behavioral changes. All the materials, including the lesson plan guidebook, were developed in conjunction with the Florida chapter of the American Cancer Society and reviewed by educators from Texas, Florida, and California.

The Florida chapter of the American Cancer Society provided training for teachers throughout Florida, as well as awareness sessions at Florida's health education conferences. In 1993, a copy of *Tobacco* was distributed to every school in Florida for their fifth-grade class, entitled "Smoke Free 2000."

Many schools are looking to health educators as the technology leaders, as a result of their training in the uses of multimedia. Health educators have been called upon to conduct technology workshops and have been involved in video conferencing. It is easy to develop multidisciplinary multimedia media programs. One program on AIDS can cover health issues related to math, science, and social studies content areas. They encourage teachers of various subjects to work together cooperatively to improve health information and behavior for today's students. Social studies and science teachers are coming to the health educator for help and advice on how multimedia can be integrated into their curriculum.

Health education has been integrated into other curriculum areas. Schools are including health education as a basic skill along with reading, writing, and arithmetic. The *ABC NEWS InterActive* health series, *Understanding Ourselves* is curriculum integrated. The guidebook provides lesson plans for health educators as well as for subjects such as social studies, math, and science, Many teachers of these subjects include health education lessons in their subject areas. The lesson in *AIDS* on "The History of Epidemics", is very popular with social studies teachers, as is "The Influence of the Tobacco Industry" on *Tobacco*.

CONCLUSION

According to Anita Hocker, Health Education Supervisor, Sarasota County Schools, Sarasota, Florida, "Integrating technology into health education provides a way to significantly foster the education goals of the 21st century with the end result of educational excellence and the highest level of student learning." (Personal communication, 1991).

The pressure is on educators to meet the challenges of today's youth and the situations they face in health education. Never in recent memory has there been such pressures on educators and publishers to meet the pressing demands of today's students and to adopt solutions to meet the current crisis in health education. The pressure is coming from the communities that rely on schools and education to provide information that once came from the home, from health professionals who view an educated youth as a vehicle to lead in efforts of prevention education. It also comes from young people, bombarded with the stimulation of television, and the opportunity to become involved in risky behaviors. They are looking for guidance and accurate information to help them make positive critical life decisions.

It is important to provide educators with powerful teaching tools to grab the attention of today's student for the precious few minutes they have. The return on this investment will pay off in the health of tomorrow's adults. We offer the following recommendations toward that goal: (a) adequate computing and networking technology in every school; (b) adequate staff development and inservice training for teachers; (c) technology training in higher education programs for students preparing for careers in education and health care; (d) evaluation and monitoring of all training programs with input from teachers, administrators, students, and families, and representatives from the community.

REFERENCES

Carlson, H. L., & Falk, D. R. (1989). Effective use of interactive videodisc instruction in understanding and implementing cooperative group learning with elementary pupils in social studies and social education. *Theory and research in social education, 17*(3), 241–258.

Carlson, H. L., & Falk, D. R. (1991). Effectiveness of interactive videodisc instructional programs in elementary teacher education. *Journal of Educational Technology Systems, 19*(2), 151–163.

Centers for Disease Control and Prevention National Center for Health Statistics (1989). *Morbidity and mortality weekly report (MMWR),* 38, 165–170.

Fletcher, J. D. (1989, January). *The potential of interactive videodisc technology for defense training and education. (Report to Congress).* Arlington, VA: Institute for Defense Analysis.

Fletcher, J. D. (1990, July). *Effectiveness and cost of interactive videodisc instruction in defense training and education* (Institute for Defense Analysis Report R2372). Arlington, VA: Institute for Defense Analysis.

Florida Vital Statistics. (1990). State of Florida Department of Health and Rehabilitation Services, Office of Vital Statistics, p. 51.

Moore, L. J., & Carnine, D. (1989). Evaluating curriculum design in the context of active teaching. *Remedial and Special Education (RASE) 10,* (4), 28–37.

Slike, S. B. (1989). The efficiency and effectiveness of an interactive videodisc system to teach sign language and vocabulary. *American Annals of the Deaf, 143,* (4), 288–290.

Telecommunications Development Center (1989). *A selected interactive videodisc bibliography* (TDC Research Report No. 2). Montgomery, Rae: Sayer, Scott.

Waters, M. (1991). Strengthening health education. *Florida Youth Risk Behavior Survey* (pp. 10–48). Tallahassee, FL: Florida Department of Education.

Transforming Health Education Via New Media

Chris Dede
Lynn Fontana
George Mason University

This chapter describes an alternative approach to health education and improving public wellness that centers both on new theories about learning and on emerging educational technologies. Network-based multimedia and distributed simulation are intriguing vehicles for health-related education rapidly becoming practical through the evolution of America's national information infrastructure (NII). We begin with a discussion of what types of learning may induce shifts in individual behavior patterns, leading to healthier lifestyles; then describe how sophisticated media can empower implementing these models of learning to promote public wellness.

IMPLICATIONS OF NEW LEARNING THEORIES FOR HEALTH EDUCATION

New ideas from cognitive science about the nature of learning are inducing educators at every level to restructure instructional environments. Although a detailed discussion of developments in learning theory and cognition is beyond the scope of this chapter, some basic principles that underlie models of learning follow here (Dede, 1990):

- Learning involves an evolutionary series of progressively more sophisticated conceptions of reality. Learners interpret every instructional experience through an existing mental model; they are not empty vessels to be filled (Bransford, Franks, Vye, & Sherwood, 1989).

- New concepts and skills are best remembered and integrated with existing knowledge if learning is active and constructive rather than passive and assimilative (Resnick, 1987).
- Learning is motivated by multiple factors, including intrinsic curiosity, social interaction, extrinsic rewards, and the joy of accomplishment. Because of this mixture of motivations, cooperative learning in groups is sometimes the optimal strategy; in other situations, individual learning is superior.

Often, the ideal environment for learning is very diverse, mixing different ages, developmental levels, and cultural backgrounds (Devillar & Faltis, 1991).

- Learning is dependent on an individual's beliefs and attitudes. For example, understanding a student's cultural perspective is vital to communicating knowledge and to measuring learning. Also, each individual has a unique style of learning, based on cognitive, sensory, psychomotor, and social factors. Tailoring instruction to this style is vital for educational effectiveness (Cohen, 1987).
- The activities through which learning takes place and the context within which learning occurs are integral parts of the knowledge gained. Educational effectiveness is enhanced by situating learning in an environment similar to that in which the knowledge will be used (Brown, Collins, & Duguid, 1989).
- Once learners have mastered the fundamentals of a subject, using an interdisciplinary approach to subject matter is more effective than breaking material into the traditional disciplines, because real-world problem solving involves comprehending richly interconnected systems (Dwyer, Ringstaff, & Sandholtz, 1991).
- Each step in learning requires time for reflective ideation. Compressing the time necessary for an individual to master an educational experience results in little or no learning. Learning is continuous and unbounded; people who treat every situation as an opportunity for growth learn more than those who limit their education to classroom settings.

These principles are centered on learning rather than teaching. For three decades, intensive efforts to improve teaching based on traditional, behaviorist models of learning-by-assimilation have had little effect on student outcomes. Increasing instructional effectiveness by focusing on cognitive science's new theories about learning-by-doing is an important shift in strategy.

Over the next decade, the challenge facing instructional designers is adapting general heuristics about new models for learning to specific educational environments delivered by advanced media such as the NII. This chapter focuses on emerging information technologies that enable sophisticated new approaches to health education. The two technologies discussed in the following sections, networked multimedia and distributed simulation, offer substantial promise of improving public knowledge about wellness and inculcating healthier behavior patterns.

MULTIMEDIA AND HIGHER-ORDER THINKING SKILLS

Research is just beginning to reveal the full power of multimedia technologies to enable individuals to develop higher-order thinking skills. The most promising outcomes of these studies focus on the amalgamation of multimedia capabilities rather than on the significance of any one attribute. Multimedia technologies can contribute to the development of higher order thinking skills because they facilitate learning via structured discovery, motivation, multiple learning styles, the navigation of web-like representations of knowledge, learner authoring of materials, the collection of rich evaluative information, and collaborative inquiry.

Because teachers in all fields face the challenge of helping students to develop basic literacy for the information age, various educational reform movements highlight higher order thinking skills as integral aspects of the new curricula. This nationwide initiative to transform teaching/learning provides an overarching framework for examining how multimedia and other emerging technologies might be used to transform health education. Although the goals of this reform movement are revolutionary, its process is likely to be evolutionary, since developers can learn from past experiences as they progress along the continuum from familiar instructional devices to more exotic and sophisticated technologies.

Using multimedia to develop thinking skills is appropriate and necessary across the curriculum including history, the natural sciences, the social sciences, and health. Design heuristics are emerging about how to create multimedia tools that give teachers leverage on difficult issues of cognition, learning styles, motivation, cooperative learning, evaluation, and the unique needs of at-risk learners. For example, the quality of multimedia databases, particularly the production values of the video, is an important factor in creating effective prototypes that promote higher-order thinking. Also, professional development activities that utilize innovative distance learning technologies such as on-line databases, teleconferences, and computer networks are central to successful utilization of multimedia products.

The amount of data available on health and wellness-related issues increases exponentially each year. This presents an ever-mounting challenge for health educators, who are attempting to help citizens assimilate information as part of their decision making and behavioral schema, not simply gain access to data. The computer has become a central device in the generation and transmission of data; now, sophisticated instructional technologies are increasingly being viewed as appropriate tools for helping learners act upon information received.

The same technologies that are currently swamping learners in data can help them master thinking skills that will promote the synthesis of knowledge. This requires a refocusing of current uses of multimedia in the curriculum, from engines for transmitting massive amounts of data to tools for structured inquiry based on higher-order thinking. Such an approach should be equally applicable to material in science, mathematics, the social sciences, and the humanities.

The most appropriate role for multimedia and emerging technologies in schools is not to augment data delivery in conventional instruction, but instead to foster a new model of teaching/learning based on learners' navigation and creation of knowledge webs through a formal inquiry process. Dede (1990) suggests that higher-order thinking skills for structured inquiry are best acquired where:

- Sophisticated information-gathering tools are used to stimulate learners to focus on testing hypotheses rather than on plotting data.
- Multiple representations for knowledge are used to help tailor content and suit individual learning styles.
- Learners construct knowledge rather than passively ingest information.
- Learning is situated in real-world contexts rather than based in artificial environments such as end-of-chapter textbook questions.
- Peers interact collaboratively, similar to the team-based approaches underlying today's science.
- Individualized instruction targets teacher intervention to assist each learner in solving current difficulties.
- Evaluation systems measure complex, higher-order skills rather than simple recall of facts.

Past Experiences with Media and Technology

To realize the potential of emerging technologies to transform health-related behavior, individual citizens must have access to these technologies, be motivated to use them, and have the skills necessary to apply experiences and information received to their own lives. Unfortunately, simple awareness of a potentially dangerous health behavior patterns usually does not lead to lifestyle changes, and motivation alone quickly fades without support systems to bridge from understanding a problem into altering unhealthy patterns of behavior.

For example, addiction to drugs, alcohol, and tobacco costs billions annually; smoking alone, now the leading single cause of death in the United States, is responsible for more that a half million deaths each year. Illness related to or complicated by poor diet and obesity adds substantially to annual health care costs. These trends toward personal pathology are reversible, if individuals choose to change their habits, but informing the public in great detail about potential risks has led to less behavioral change than hoped.

Using new technologies to present still more information on these issues will not provide much additional leverage. Significant changes in behaviors that lead to healthier individuals require the design of information systems that personalize wellness information, emphasize individual responsibility, and promote the internalization of knowledge and its immediate application in daily life. As long as information about health choices remains noninteractive, impersonal, and moralistic, improvements in citizen's decisions are unlikely.

Clearly, new information technologies are tools with which to invent solutions to these challenges. Emerging national information infrastructure services enable users to conduct individualized, confidential risk assessments; confer with experts on the latest in medical research; access the most up-to-date information on treatment and medication; and utilize customized tools to explore personal health care options. These capabilities will only be valuable, however, if individuals use them. Consumers need to be motivated to seek information and to convert that data into knowledge that guides healthful decisions.

To assure the effectiveness of multimedia, we must use the full power of this educational delivery system, exploiting the unique strengths of each facet. In particular, visual materials can capture the public's interest, draw them into a subject so that they will want more information, and promote a search for personal meaning. Visuals, particularly full-motion video, can interest learners in topics and make them ready and willing to act on that information in constructive ways. Wellness-based educational approaches will be most successful if they engage emotions as the first step in attaining a new intellectual perspective.

As an example, think about the impact of Ken Burns' public broadcasting system video production work, *The Civil War*, on viewers' desire to know more about this event in history. As a result of this video series, millions of copies of the videocassette and accompanying book were sold. The "Civil War" sections of bookstores were laid bare within weeks. Because of the demand, publishers were prompted to reprint dozens of books on the topic. In as simple a medium as broadcast television, a well-designed program deeply touched an audience and whetted its appetite for more information about an event that transpired over a century ago.

How might such a model of instructional design be applied in health care? In the mid-1980s, a similar epiphany, albeit smaller and less publicized, occurred in the health and safety arena. A Washington, DC-based news reporter, Kelly Burke, produced a public broadcasting special entitled *Drinking and Driving: The Toll and the Tears*. This special was launched in partnership with major health education groups concerned about driving while intoxicated. At the time, this strategy of using the broadcast medium with an accompanying discussion guide as the centerpiece for an outreach effort, in collaboration with national groups, was state-of-the art. When this television special first aired on PBS in prime time, two million viewers were reached, but this was only the beginning of a chain reaction that touched and positively influenced several million more.

The impact of the program was so great that the toll-free number offering the discussion guide was clogged for weeks. Public broadcasting stations around the nation re-aired the program dozens of times in response to local requests. Thousands of copies of the cassettes and discussion guides were distributed nationally. The program was licensed to instructional television for distribution to thousands of schools all across the country.

How did the design of this effort make it so effective? A high quality television program was crucial; without being preachy, the effectively organized, dramatic information grabbed viewers' attention. The documentary presented a series of personal vignettes of people who, while under the influence of alcohol, had been responsible for automobile accidents that took the lives of their victims. In the final vignette, the program's producer Kelly Burke revealed his own story. After having only two drinks, Kelly had caused an accident that resulted in the death of a local off-duty police officer. The entire documentary project had been part of his court-imposed sentence. This revelation had a significant emotional impact on viewers; ignoring the realities of the drinking and driving was impossible for those who realized how easy it would be to end up in Kelly's shoes.

The televised program was only the beginning. Health educators were part of the team that developed and distributed the guides and cassettes and promoted the broadcast. These materials were designed as a tool that helped health service providers to do their jobs more effectively. The video/print package was used in thousands of community groups, classes in schools, and courses for convicted DWI (driving while intoxicated) offenders.

The emphasis was on the individual's need to take responsibility, and the materials outlined clear steps to making a difference. Although no study was done to prove that the program and guide were directly responsible for a decline in drinking and driving incidents, sufficient anecdotal evidence has convinced many in the media and in wellness fields that these materials empowered health professionals to make a significant impact.

What does this have to do with emerging learning environments? Several design heuristics seem evident:

- Touching the emotions of individuals is the first step in gaining their attention and preparing them for a learning experience and eventual action.
- The development of new systems that use a variety of technologies must be done by teams of professionals that include video producers, graphic artists, computer programmers, content specialists, and instructional designers.
- The design lifecycle necessitates an interactive process with both representatives of the intended audience and with health educators.

The central challenge is to understand how learners use various forms of interactive technologies to transform external information into personal knowledge. As one illustration of work in progress with "lessons learned" for health education, at George Mason University a project in development looks directly at how multimedia can be designed to promote higher order thinking skills in history.

Multimedia Design Elements

In the Multimedia and Thinking Skills Project, we have focused on five elements to be addressed in the design of effective instructional multimedia products: explicit instruction, modeling, tutoring/coaching, learner control, and collaborative learning. These design elements led us to create five features of our multimedia system: the *IBI (Inquiry Bureau of Investigation), Guided Tours, Dr. Know, Custom Tours,* and the *Production Console.* (Other design elements such as the quality of the database, equity, and professional development are contextual factors that also must be considered if successful utilization of multimedia is to be accomplished, but are not detailed here for reasons of space.)

The *IBI* is literally an iconic bureau; by opening each of its drawers, learners receive explicit instruction on the steps of inquiry. *Guided Tours* model the inquiry process by taking learners through the database on carefully designed paths that pose questions and compel them to evaluate the extent to which the data they encounter helps to answer these questions. *Dr. Know* is the context-sensitive coach or tutor who helps learners develop data-gathering and metacognitive skills. *Custom Tours* is a non-structured access system to the data that allows free-form searches based on learners' interests and in their control. The *Production Console* gives learners the tools with which to manipulate the information in the database to make reports and create their own tours; this not only enables learner control but also facilitates collaborative learning.

Explicit Instruction: Inquiry Bureau of Investigation (IBI)

Accessing the *IBI* provides learners with explicit instruction in the steps of inquiry. Because higher-order thinking is neither instinctive nor developed as a result of teaching only content, providing opportunities in subject matter instruction for learners to use thinking skills is not enough (Feuerstein, 1980; Perkins, 1987; Sternberg, 1984). Instead, learners need continuing, deliberate, and explicit instruction in how to use inquiry skills not in a decontextualized manner, but simultaneous with their striving to master subject matter (Salomon, 1991; Whimbey & Whimbey, 1975). Additional studies of database use in social studies (Ehman, Glenn, Johnson, & White, 1992; White, 1987) and hypertext use in science education (White, 1989) support the view that effective use of information by learners requires an array of thinking skill supports. In a study of word processing as a way to improve writing, Zellermayer, Salomon, Globerson, and Givon (1991) indicate that courseware with guided instruction and continuous high-level cognitive help contributed to substantial improvement sustained over time and across technologies.

Modeling: Guided Tours

Guided Tours model historic and scientific inquiry. They are purposeful environments that deliberately present contradictory observations, reports, or accounts, inspiring learners to invent additional questions, hypothesize answers, and identify potential sources of data. We have incorporated into our design an anthropomorphic coach, *Dr. Know,* who acts in *Guided Tours* to facilitate learners developing inquiry skills.

As Scardamalia, Bereiter, McLean, Swallow, and Woodruff (1989) discuss, learners require extensive scaffolding to sustain active, constructive approaches to learning. Their approach centers on procedural facilitation of learning not machine-based intelligence, but implicit structure and cognitive tools that enable learners to maximize their own intelligence and knowledge. In the context of multimedia, procedural facilitation requires the instructional system to present exemplary knowledge-structuring architectures, illustrative representational formats, and prompts that encourage expert information-managing strategies. These are not only powerful mechanisms for inducing learning, but also inculcate core skills for communications literacy in tomorrow's workplace (Sproull & Kiesler, 1991). In our design, the *Guided Tours* provide the first two of these strategies for enhancing learning, and *Dr. Know* provides the third.

Tutoring/Coaching: Dr. Know

In our prototypes, a resident scholar or context-sensitive coach (*Dr. Know*) can be invoked to assist learners in developing data gathering and metacognitive skills at each stop along the *Tour*. Laurel, Oren, and Don (1990) describe a variety of ways that interface agents serving as guides can reduce the cognitive load for users of multimedia systems. In *Guided Tours, Dr. Know* gives context-sensitive advice on how to explore a cluster of data, how to think through issues in an inquiry situation, and what questions may be most productive to pursue. In the *IBI, Dr. Know* gives systematic instruction on the steps of inquiry within the context of the tour. Instruction about thinking skills at the time that they are contextually needed to achieve subject matter mastery motivates learners to acquire these skills and enhances the quality of subsequent learning. For example, White (1987, 1989) notes the importance of the interactive coach in supporting skill development.

In many ways, we are wrestling with the same issues as the Cognition and Technology Group at Vanderbilt (1990), who are examining the concepts of anchored instruction and situated cognition. They are creating instructional systems for various domains that permit learners and teachers to experience inquiry through the actions of an expert. Our *Dr. Know* is an expert scientist or historian who takes learners (whom the Vanderbilt Group might describe as apprentices) along on an investigation. Through modeling, *Dr. Know* helps learners develop their skills and knowledge. This investigative setting for anchoring instruction in an authentic context enhances learner motivation and makes the skills more likely to be generalized into real-world environments.

For subject-independent material such as the *IBI, Dr. Know's* help transfers across disciplinary domains. For specific *Tours* in a particular subject, this context-sensitive help must be customized from generic templates by the instructional design team creating the multimedia materials for that domain. *Dr. Know* does not have generative capabilities, so is not an intelligent coach; the complexity of developing a discipline-independent, glass-box, knowledge-based system would exceed the benefits such a feature could provide. Because the inquiry model is generalizable across disciplines, our preliminary work indicates that canonical variations of prompts make

Dr. Know easily adaptable to a wide range of subject domains. Unlike the group at Vanderbilt, we have incorporated explicit as well implicit instruction in inquiry.

Learner Control: Custom Tours

The environment we are developing supports instruction in which learners have some freedom of choice including producing their own reflective multimedia materials as an outcome measure indicating mastery and in which they work in small groups using collaborative inquiry to explore and master multimedia material. In *Custom Tours*, learners can use the facilities of the *Production Console* (described in the following section) to create their own paths through the data.

Once learners feel they know the inquiry process, they are encouraged to design their own *Custom Tours*. As learners take more responsibility for their own intellectual exploration, *Dr. Know* prompts them to state "Thoughtful Questions and Hypotheses" that will help them establish what Scardamalia et al. (1989) term goal orientation. A major objective of the goal orientation strategies Dr. Know conveys is to de-emphasize the tendency of naive learners (and teachers) to rehearse or memorize information; instead, stress is placed on skills that allow data to be reconfigured as needed.

Helgeson (1987) establishes the importance of learner freedom, choice, and the use of small groups as effective in improving scientific problem-solving behaviors. However, whereas freedom for learners to navigate and explore knowledge webs is important, our prototypes will also provide extensive implicit structure via *Guided Tours* through the multimedia database. As Rivers & Vockell (1987) indicate in their study of computerized simulations and problem-solving skills, learners using guided versions of simulations surpassed those using unguided versions on tests measuring cognitive processes and critical thinking.

Learner Control: The Production Console

Beyond the cluster of capabilities associated with *Dr. Know* and the *IBI*, the multimedia/thinking skills shell enables learners to produce their own multimedia presentation and *Custom Tours*. The *Production Console* is a tool to help learners construct both *Custom Tours* built around their own Thoughtful Questions and Reports that document their proof/disproof of the hypothesis from a *Guided Tour*. In the *Production Console*, learners manipulate the data in various forms of video, audio, graphic, animation, and text that they have gathered in their Journal. Learners can add additional data external to the prototype, can analyze their Journal entries based on the inquiry process in the *IBI*, and can author their own *Custom Tours* to augment those provided with the instructional system.

Participation in audiovisual production contributes to student learning in curriculum areas such as language arts, social studies, and science. Learners engaged in these production activities demonstrated improvements in self-concept, motivation, creativity, and attitude. In addition, these learners increased their involvement with print media (Donoho, 1986). The *Production Console* is important for another reason: Learners use this feature to demonstrate their knowledge, their ability to analyze data and state conclusions, and their skills in applying what they have learned. An objective of our project is to develop improved evaluative measures that can help teachers assess learners' mastery of inquiry skills.

Cooperative Learning

Capabilities such as the *Production Console* support cooperative learning because learners can easily share and display data, comment on drafts, and keep collective records of their inquiry process. The production of multimedia tours is a particularly powerful vehicle for collaboration because learners must pool unique individual strengths (editing video, scanning images, digitalizing sound, creating animation, writing text) to develop a successful product. Although there has been limited research on collaborative or cooperative learning in multimedia environments, Adams, Carlson, & Hamm's (1990) review of the research indicates that collaborative learning with interactive technologies leads to higher scores on measures of content knowledge and observation skills, as well as high degrees of learner motivation and satisfaction.

Generalization to Health Education

In summary, the Multimedia and Thinking Skills Project views learners as tourists through multimedia databases. Just as tourists determine their direction and make choices about how they will explore different sights, users of our multimedia/thinking skills shell decide how to explore their cognitive environment. They may choose one of several *Guided Tours*, or they can explore the database via *Custom Tours*. A reflective context for learning is created in the exemplary *Guided Tours*; this provides a stimulating environment within which learners can become actively engaged in learning the subject domain while receiving context-sensitive instruction on the inquiry process.

As learners proceed through *Guided* or *Custom Tours,* they are able to call up their on-line tutor, *Dr. Know*, to help them evaluate information. During their tours they collect information in their electronic Journal which they take to the *Production Console*. Here they can consult the *IBI* and make certain they have followed all the steps of the inquiry process; they are then ready to create reports or design their own tours (Fontana, Dede, White, & Cates, 1993).

Using multimedia to promote higher-order thinking about health is an obvious way to employ this multimedia/thinking skills shell. The quantity of health-related information in various forms—including text, graphics, and full-motion video—is immense. As the NII becomes a reality, these databases, which reside in digital form on hundreds of servers nationwide, will be indexed and made accessible to users. The challenge facing health educators is to work with teams of instructional designers to make such databases truly usable by various audiences.

By employing the multimedia and thinking skills design elements described previously, health educators can develop applications that help learners navigate and manipulate these databases to create personal knowledge. Design principles that underlie these applications should: (a) promote interest in the topic of health by touching learners' emotions, (b) draw learners into inquiries based on their personal interests, (c) convey the inquiry and decision-making skills necessary to making good health choices, and (d) incorporate tools that enable learners to create their own paths through information and to present their own points of view

Health care multimedia applications build learners' knowledge about wellness. A complementary type of educational application, based on distributed simulation, can give learners the chance to experiment with potential lifestyle shifts based on that new knowledge. A potential, immersive "synthetic environment" for this type of learning is described in the following section.

A VIRTUAL ENVIRONMENT TO ENHANCE WELLNESS

The next section of this chapter sketches the potential evolution of a new type of learning medium for public wellness, illustrated by a hypothetical virtual environment termed *HealthWorld*. Distributed simulation, the emerging technology foundational to creating *HealthWorld*, is described; and precursors of such learning environments that exemplify their challenges and opportunities are discussed. The *HealthWorld* example is representative of a large class of new media for learning made possible by high performance computing and wide-area, high-bandwidth telecommunications, the core of America's emerging NII.

HealthWorld would be a shared synthetic environment, accessed via telecommunications, whose (un)real estate would provide common ground for learning preventive-care behaviors centered on wellness. Participants separated by distance could navigate through a shared context in *HealthWorld*, collaboratively interacting with a simulated environment that mirrors many aspects of our physical world, but also enables almost magical activities. Precollege pupils, university students, pre- and in-service health professionals, and the general public could experience virtual counseling sessions, engage in vicarious experiments with alternative lifestyles, and gain physiological and medical knowledge during shared adventures through simulations of the human body.

What Could Learners Experience in *HealthWorld*?

Via a personal computer and modem, learners would access *HealthWorld* through a cartoon-like interface, as illustrated in Fig. 9.1 (Moshell & Hughes, 1994).

As shown, on one portion of the computer monitor a user would see a graphical representation of his or her current location in the synthetic environment: a context with various objects depicted, including an "avatar" (an image of the user's persona in the virtual world). Other parts of the

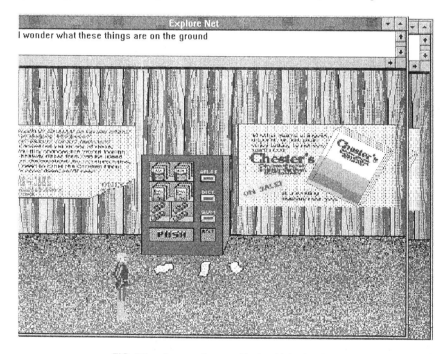

FIG. 9.1.   A scene from an ExploreNet adventure.

monitor screen would convey menus of possible actions, a dramatic narrative guiding the evolution of the learning situation, and ongoing conversations with other avatars in the virtual environment.

HealthWorld could contain at least the following types of experiential learning environments:

- Collaborative explorations through various systems of the human body, based on adventures involving drama and fantasy. This will enable depictions of disease processes and other threats to health, as well as illustrating how wellness behaviors remediate these situations.
- Virtual counseling groups in which peers can share ideas, receive emotional support, and obtain personalized medical advice while remaining anonymous.
- Mortality environments in which learners can live alternative lifestyles and vicariously experience the consequences of different behavior patterns.
- Mechanisms through which users can access additional information, either by navigating through a database or via instructing a machine-based agent to automatically scan on-line wellness materials as they are updated.

As a *HealthWorld* environment evolved, users themselves would aid in designing other types of virtual settings and activities.

The Power of Virtual Environments for Health-Related Learning

As discussed earlier in the section on cognitive science research, studies indicate that learning-by-doing may enable greater change in health-related lifestyle behavior patterns than learning-by-assimilating. Experiential learning is highly motivating, promotes greater retention of content, and conveys behavioral skills in addition to conceptual comprehension. Moreover, since participants self-generate much of the educational situation through shared interaction, collaborative simulations require less development funding than presentational materials. In contrast, disseminating preventive-care information in a presentational format even via America's emerging high-tech communications infrastructure is unlikely to show dramatic gains beyond current strategies for transmitting public health advice. Unfortunately, even expertly done "teaching by telling" is relatively ineffective in promoting sustained changes in lifestyle.

In particular, a *HealthWorld* virtual environment would provide unique leverage on factors known to be crucial for wellness: (a) learning-on-demand in familiar settings such as the home; (b) the ability to receive personalized advice while remaining anonymous; (c) support from peers confronting similar challenges; and (d) defusing denial in potentially frightening situations via humor, curiosity, and fantasy.

In addition, because learners actively explore why threats to health are minimized by certain types of behaviors, *HealthWorld* would empower its participants to assume responsibility for their lifestyles based on personal knowledge, as opposed to following prescriptive advice based on blind trust.

Essential to the successful resolution of the U.S. health care crisis is a shift to wellness behaviors on the part of the general public. This is the least politically controversial means of capping health care costs, as well as potentially the most effective. However, to alter the primary focus of medicine in America from therapeutic intervention to preventive anticipation is an enormous task that will require large-scale, inexpensive educational techniques to encourage changes in personal lifestyle. Distributed simulation is a promising new approach whose development costs are very low relative to the potential savings in medical expenditures even a mildly successful *HealthWorld* could generate.

In addition, knowledge about health is an excellent initial domain to explore the generalizability of distributed synthetic environments to learning other types of subject matter. Preventive-care material encompasses technical concepts, psychosocial issues, ethical concerns, and procedural skills; wellness topics are also of considerable interest to a wide segment of the American population. The creation of a *HealthWorld* environment would generate design heuristics for developing this new educational delivery mechanism's full potential across a spectrum of potential applications.

The New Media Underlying Distributed Synthetic Environments

Distributed simulation is a powerful educational delivery mechanism developed by the U.S. Department of Defense in the late 1980s. For example, *SimNet* (Orlansky & Thorp, 1991) is a training application that creates a virtual battlefield on which learners at remote sites can simultaneously develop both individual and collective military skills. Complex data-objects that indicate changes in the state of each piece of equipment are exchanged via a telephone network interconnecting the training workstations ("dial-a-war"). The appearance and capabilities of graphically represented military equipment alter second-by-second as the virtual battle evolves. The fundamental capability distributed simulation enables is *telepresence*, the perception by geographically separated learners of simultaneous presence in a virtual environment. Through telepresence, a widely distributed group of personnel can engage in simulated real-time warfare without the necessity of gathering the participants at a single site to conduct combat.

This medium is becoming available for civilian sector usage via defense conversion funding and public/private support for a national information infrastructure. Such a learning environment has many possible applications in health-related education (e.g., virtual hospitals for medical training). Distributed simulation enhances students' ability to apply abstract knowledge by situating education in virtual contexts similar to the environments in which learners' skills will be used.

When the material involved has psychosocial as well as intellectual dimensions, the design of authentic experiences to embed in virtual environments for education becomes more complex. By using sophisticated telecommunications, learners can interact in rich machine-mediated psychosocial environments populated both by video-links to other people and by simulated beings. As the vignette earlier illustrates, these simulated beings may be *knowbots* (machine-based agents) or *avatars* (computer graphics representations of people); each adds an important dimension to education in synthetic environments.

One example of a training application that incorporates machine-based agents involves software engineering education (Stevens, 1989). Using hypermedia, digital video interactive (DVI), and rule-based expert systems, the Advanced Learning Technologies Project at Carnegie Mellon University has created a virtual environment similar to a typical corporate setting. The trainee interacts with this artificial reality in the role of a just-hired software engineer still learning the profession. Through direct instruction and simulated experience, students are trained in a technical process, code inspection, that is one stage of a formal methodology for software development.

The learner can access various rooms in the virtual software company, including an auditorium, library, office, training center, and conference facility. Machine-based agents (knowbots) that simulate people, such as a trainer and a librarian, facilitate the use of resources to learn about the code inspection process. Via specialized tools in the office, the student can prepare for a simulated code inspection, in which he or she can choose to play any of three roles out of the four roles possible in this formal software review process. For each inspection, a rule-based expert system utilizes DVI technology to construct knowbots that simulate the three roles not chosen by the learner. This knowledge-based system controls the topic of conversation; determines who should speak next; and models the personalities of the knowbots in the inspection meeting, altering their cognitive and affective perspective depending on what is happening.

The learner uses a menu-based natural language interface to interact with these simulated beings, who model behaviors typical in code inspection situations. The student not only can choose from a wide range of options of what to say, but can determine when to make remarks and can select the emotional inflection of his or her utterances, from a calm passive tone to an angry snarl. By mimicking the reactions likely from human participants in a real simulation, the knowbots provide the learner with a sense of the strengths and weaknesses of different intellectual/psychosocial strategies for that role in a code inspection.

Without using artificial realities and knowbots, this type of authentic experience is very difficult to simulate. Not only is the training environment often dissimilar from the context in which skills will be used; but also students do not know how to roleplay exemplary, typical, and problematic situations. Through knowbots, the instructional designer can provide paradigmatic illustrations of how to handle a variety of issues, without the expense of having teams of human actors perform for each individual learner.

This pedagogical approach could generalize to many types of medical education, such as training nurses how to handle certain kinds of patient personalities. In addition, machine-based agents could serve as guides to various environments within *HealthWorld*. Early research into educational interfaces based on conversation with human-like entities indicates that this strategy enhances learning along a variety of dimensions (Laurel et al., 1990). Knowbots could even become "surrogate participants" within the environment, interacting in ways that stimulated learning (e.g., feigning ignorance so that a user could engage in peer teaching, thereby mastering the material more deeply).

As a complement to responding to knowbots as if they were human, participants in a virtual world interacting via avatars tend to treat each other as imaginary beings. An intriguing example of this phenomenon is documented in research on Lucasfilm's *Habitat* (Morningstar & Farmer, 1991). *Habitat* was initially designed to be an on-line entertainment medium in which people could meet in a virtual environment to play adventure games. Users, however, extended the system into a full-fledged virtual community with a unique culture; rather than playing prescribed fantasy games, they focused on creating new lifestyles and utopian societies.

As an entertainment-oriented cyberspace, *Habitat* provided participants the opportunity to get married or divorced (without real-world repercussions), start businesses (without risking money), found religions (without real-world persecution), murder other's avatars (without moral qualms), and tailor the appearance of one's own avatar to assume a range of personal identities (e.g. movie star, dragon). Just as *SimNet* enables virtual battles, *Habitat* and its successors empower users to create artificial societies. What people want from these virtual societies that the real world cannot offer is magic, such as the gender-alteration machine (Change-o-matic) that was one of the most popular devices in the *Habitat* world.

Users learned more about their innermost needs and desires by participating in Habitat than they would have by spending an equivalent amount of time listening to psychology lectures. Similarly, social scientists are discovering more about utopias by studying *Habitat*'s successors than they did by researching communes, which were too restricted by real-world considerations to meaningfully mirror people's visions of ideal communities. Giving users nearly magical powers opens up learning in ways that educators are just beginning to understand. As with any emerging medium, first traditional types of content are ported to the new channel; then alternative, unique forms of expression like *Habitat* are created to take advantage of expanded capabilities for communication and education.

Whether or not distributed simulation environments for public wellness resemble the hypothetical *HealthWorld* we have depicted, this type of emerging technology will have a profound impact on health education. The NII provides a huge, ubiquitous channel along which new kinds of learning experiences will travel. Multimedia and virtual environments have complementary strengths; their usage and potential fusion will open up extraordinary vistas for health policy and practice.

CONCLUSION

Schuler (1994) described how the inexpensive availability of personal computers and low bandwidth telecommunications is fostering the rapid development of multiple, community-oriented electronic bulletin boards and networks. Often, the focus of these community networks is on developing opportunities for disadvantaged groups and encouraging widespread participation in democratic political processes. As the NII evolves, low-cost access to high-performance computing and wide-bandwidth telecommunications will allow these community networks to implement educational services based on sophisticated technologies such as multimedia and distributed simulation. Health is a predominant concern for many citizens, so information and education about wellness-related issues will find community networks a ready audience already adept at using these media. These groups are potentially a powerful complement to existing mechanisms for disseminating health-related education.

The new vehicles for dissemination that technologies such as networked multimedia and distributed simulation provide is analogous to a virtual "theater" in which potential shifts in individual lifestyle can be studied and explored. Participatory computer-based environments such as those described in this chapter are powerful vehicles for mimesis (Laurel, 1991).

Although this can lead to escapism, as a dramatist knows, with good design the focus of mimesis shifts to playful exploration, learning by doing, and catharsis: all important processes for inducing wellness behaviors. As one example of this theatrical approach to education, the Multi-User Shared Environments (MUSEs) now appearing on national computer networks are a new form of networked educational application that focus on shared learning within a virtual, text-based world (Rheingold, 1993).

The key capability that MUSEs, learner-constructed multimedia, and distributed simulation add to current educational technologies is immersion (the subjective impression that a user is participating in a "world" comprehensive and realistic enough to induce the willing suspension of disbelief). The induction of immersion depends in part on both actional and symbolic factors. Inducing actional immersion involves empowering the participant in a virtual environment to initiate actions that have novel, intriguing consequences. For example, when a baby is learning to walk, the degree of concentration this activity creates in the infant is extraordinary. Discovering new capabilities to shape one's environment is highly motivating and sharply focuses attention.

Inducing a participant's symbolic immersion involves activating powerful semantic associations via the content of a virtual environment. As an illustration, reading a horror novel at midnight in a strange house builds a mounting sense of terror even though one's physical context is unchanging and rationally safe. Invoking intellectual, emotional, and normative archetypes deepens one's experience in a virtual environment by imposing an complex overlay of associative mental models.

Research that leads to a better understanding of how immersion and telepresence empower learning is central to improving the type of health education discussed in this chapter. Moreover, in the long run, networked educational environments may empower more than wellness behaviors. As America's NII develops, Rheingold (1993) portrayed a fundamental choice between a virtual forum that enables true democracy and open-ended learning ("Athens without slaves") or a pervasive surveillance medium for propaganda and escapism (virtual "bread and circuses"). Creating multimedia and distributed simulation applications that encourage the former outcome is an important means for ensuring full public access to and control of the new media.

REFERENCES

Adams, D., Carlson, H., & Hamm, M. (1990). *Cooperative learning & educational media*. Englewood Cliffs, NJ: Educational Technology Publications.

Bransford, J. D., Franks, J. J., Vye, N. J., & Sherwood, R. D. (1989). New approaches to instruction: Because wisdom can't be told. In S. Vosniadou and A. Orton (Eds.), *Similarity and Analogical Reasoning* (pp. 111–145). New York: Cambridge University Press.

Brown, J. S., Collins, A., & Duguid, P. (1989). Situated cognition and the culture of learning. *Educational Researcher, 18*(1), 32–42.

Cognition and Technology Group at Vanderbilt (1990). Anchored instruction and its relationship to situated cognition, *Educational Researcher, 19*(3), 2–10.

Cohen, D. K. (1987). Teaching practice: Plus ca change . . . In P. Jackson (Ed.), *Contributing to educational change: Perspectives on research and practice* (pp. 76–89). Berkeley, CA: McCutchan.

Dede, C. (1990). Imaging technology's role in restructuring for learning. In K. Sheingold & M. S. Tucker (Eds.), *Restructuring for learning with technology* (pp. 49–72). New York: Center for Technology in Education, Bank Street College of Education and National Center on Education and the Economy.

Devillar, R. A., & Faltis, C. J. (1991). *Computers and cultural diversity: Restructuring for school success.* Albany, NY: State University of New York Press.

Donoho, G. (1986). *Measure of audiovisual production activities with students.* Philadelphia, PA: College of Information Studies, Drexel University.

Dwyer, D. C., Ringstaff, C., & Sandholtz, J. H. (1991). Changes in teachers' beliefs and practices in technology-rich classrooms. *Educational Leadership, 48*(9), 45–52.

Ehman, L. H., Glenn, A., Johnson, V., & White, C. S. (1992). Using computer databases in student problem solving: A study of eight social studies teachers' classrooms. *Theory and Research in Social Education, 20*(2), 179–206.

Feuerstein, R. (1980). *Instrument enrichment.* Baltimore: University Park Press.

Fontana, L., Dede, C., White, C. S., & Cates, W. M. (1993). Multimedia: A gateway to higher order thinking skills. In M. R. Simonson & K. Abu-Omar (Eds.), *Fifteenth Annual Proceedings of Selected Research and Development Presentations, National Convention of the Association for Educational Communications and Technology* (pp. 351–364). Washington, DC: AECT.

Helgeson, S. L. (1987). *The relationship between curriculum and instruction and problem solving in middle/junior high school science.* Columbus, OH: The Ohio State University, SMEAC Information Center. (ERIC Document Reproduction Service No. ED 290 606)

Laurel, B. (1991). *Computers as theater.* Menlo Park, CA: Addison-Wesley.

Laurel, B., Oren, T., & Don, A. (1990). Issues in multimedia interface design: Media integration and interface agents. *Proceedings of the ACM computer-human interface conference* (pp. 133–139). New York: Association for Computing Machinery.

Morningstar, C., & Farmer, F. R. (1991). The lessons of Lucas film's Habitat. In M. Benedikt (Ed.), *Cyberspace: First steps* (pp. 273–302). Cambridge, MA: MIT Press.

Moshell, J. M., & Hughes, C. E. (1994). Shared virtual worlds for education. *Virtual Reality World, 1*(2), 63–74.

Orlansky, J., & Thorp, J. (1991). SIMNET an engagement training system for tactical warfare. *Journal of Defense Research, 20*(2), 774–783.

Perkins, D. (1987). Introduction. In Barry K. Beyer (Ed.), *Practical strategies for teaching thinking* (pp. xi–xiv). Boston: Allyn & Bacon.

Resnick, L. B. (1987). Learning in school and out. *Educational Researcher, 16*(9), 13–20.

Rheingold, H. (1993). *The virtual community: Homesteading on the electronic frontier.* New York: Addison-Wesley.

Rivers, R., & Vockell, E. (1987). Computer simulations to stimulate scientific problem solving. *Journal of Research in Science Teaching, 24*(5), 403–415.

Salomon, G., Perkins, D. N., & Globerson, T. (1991). Partners in cognition: Extending human intelligence with intelligent technologies. *Educational Researcher, 20*(3), 2–9.

Scardamalia, M., Bereiter, C., McLean, R., Swallow, J., & Woodruff, E. (1989). Computer-supported intentional learning environments. *Journal of Educational Computing Research, 5*(1), 51–68.

Schuler, D. (1994). Community networks: Building a new participatory medium. *Communications of the ACM, 37*(1), 39–51.

Sproull, S., & Kiesler, S. (1991). *Connections: New ways of working in the networked world.* Cambridge, MA: MIT Press.

Sternberg, R. (1984). How can we teach intelligence? *Educational Leadership, 42*(1), 38–48.

Stevens, S. (1989). Intelligent interactive video simulation of a code inspection. *Communications of the ACM, 32*(7), 832–843.

Whimbey, A., & Whimbey, L. S. (1975). *Intelligence can be taught.* New York: E. P. Dutton.

White, C. S. (1987). Developing information processing skills through structured activities with a computerized file-management program. *Journal of Educational Computing Research, 3*(3), 355–357.

White, C. S. (1989). *A field test of the hypertext product "Scientists at Work": Report of preliminary findings* (Paper presented at the National Educational Computing Conference, Boston, MA). Fairfax, VA: Graduate School of Education, George Mason University.

Zellermayer, M., Salomon, G., Globerson, T., & Givon, H. (1991). Enhancing writing related metacognition through a computerized writing paper. *American Educational Research Journal, 28*(2), 373–391.

10 Meditation on the New Media and Professional Education

Joseph V. Henderson
Dartmouth Medical School

Marshall McLuhan said a great deal in his life, some of it overblown rhetoric, much of it profound. In talking about "new media" he emphasized that we need to be careful about using them, that we urgently need to understand them and how they work. The problem is that new media can be so powerful and all-encompassing that they overwhelm. Every new medium creates its own environment, which can affect our perceptions in a "total and ruthless" fashion. The new medium does not just add itself to what already exists; it transforms it. The computer, for example, has to be viewed in the context of the data networks, the methods of communicating and viewing information, the reduction of the world to that which can be conveniently represented in binary form—the all-encompassing changed habits and situations it has brought with it.

McLuhan was not speaking of the new media as in this book, but any extension of ourselves that is novel and is adopted, the automobile as much as the telephone and television. However, he foresaw a good deal of what we can include in our current discussions of new media, in particular world-spanning communications networks (firmly establishing his "global village")[1] and personal, portable computers that help us mesh our personal experience with the experience of that "great wired brain of the outer world."

Today we can add what many are calling interactive multimedia, combining computers and media (text, still and moving images, sound), yielding tools that can focus our perception and facilitate understanding. And further

---

[1] Actually first described in 1934, along with many other germane insights, by Lewis Mumford in his *Technics and Civilization,* New York: Harcourt, Brace, and Co.

along the same continuum, but with the same purpose, lies virtual reality, providing a visual (ultimately audible and tactile) sense of immersion in a three-dimensional environment that is completely computer-generated.

This chapter discusses the new media and how they might be used in the education of health professionals, written from the dual perspectives of an educator who designs, produces, and uses new media programs and of an epidemiologist who has wrestled with the issues of building a large, multimedia database. The chapter contains views that are more personal and philosophical, and less technical or research-based. McLuhan pointed out that the effects of a new medium are powerful, but they are usually so pervasive that they are insensible. The effects of the new media are already powerful and pervasive, and they will grow much more so. My goal is to help make them a bit more sensible.[2]

First, a précis establishes a broad context for considering new media in health education. Following this, a brief critique summarizes some current thinking on medical education as it relates to health care reform; there is an emphasis on the education of generalist care providers. Finally, a more specific discussion, with brief diatribe, considers the two predominant—and countervailing—approaches to using new media in education: drill and practice for rote memorization of facts on the one hand and, on the other, unguided, unstructured exploration and discovery meanderings through collections of multimedia information. A third alternative is proposed, based on the ideas of Dewey and Schön; I advocate use of new media to provide guided experiences and reflective practicums that can help professionals integrate theory-based "facts" with clinical practice, to help them learn the "art" of providing care, and that can help prepare them for the real, "swampy" world of professional practice.

CONTEXT: THE INFORMATION (KNOWLEDGE) AGE

As you've undoubtedly heard, we're entering what many call the information age. Evidence for this abounds: Headlines, speeches, and no small amount of hype declare the imminent arrival of information superhighways; 500-channel, video-on-demand, interactive TV set-top boxes; interactive multimedia, and so forth. A national information infrastructure, capable of

---

[2]A note to the reader concerning my writing style. This chapter is intended to be, as its title suggests, a "meditiation," a mulling over, a contemplation of the playing field. I find this style to be the one most consistent with this goal. Were I offering data, summaries of experiences in testing, reviews of the literature or the experience of others in using new media, a more academic tone would be appropriate. I'm striving to provide a readable, reasonably thought-provoking commentary in the spirit of other critics of technology such as Neil Postman and McLuhan himself.

transferring data at rates in the gigabits per second range is starting really to be built, funded through an amalgam of public and private sources. CD-ROM-based interactive computer entertainment, educational, and informational programs are starting to roll off the presses, sales of CD-ROM players are soaring and publishers are scrambling to define and take advantage of the immediate market for these new electronic media—and, more long range, to develop programming and publishing paradigms to be ready for this superhighway they've been hearing about.

Business leaders and pundits are more and more talking about a new world economy based on information as a main product (even when hard goods are involved), accompanied by major reorganization and relocation of workers; telecommuting will become the norm, they predict, with central offices more a club-like setting for social contact; workers will leave cities and move in increasing numbers to rural communities.[3] Leaders in health care are starting to regard health information as a useful and marketable product of the care system, and to advocate the use of new media to support decisions made by recipients of care as well as providers.

So, we're leaving the Industrial Age and entering the Information Age or, as some prefer it, the Knowledge Age (assuming that information is only of benefit if you know something useful after you've had a chance to mull things over). In fact some think in terms of a hierarchy with data at the bottom, somehow being processed into information, then into knowledge. Given current and emerging technologies, data can be put together in fairly automatic ways, with bits of data correlated and packaged for viewing. To get to knowledge, people usually have to get involved and start thinking about what these correlated bits mean, and various technologies can assist that, from statistics and simple tables and graphs to complex, multidimensional, scientific visualization, computer displays. Others say that the knowing comes in the application of information, usually in actual practice, but more and more in computer-based simulations.

This leads to another step in this hierarchy, moving from having knowledge to achieving wisdom: a more difficult step. Wisdom involves bringing the perspective of personal experience and applying knowledge. Not in an arms-length and abstract way, but with a deeper understanding of how that application can affect our world and the people in it (however we define that world: our practice, our region, or our nation). Technology *might* be used to help us bridge the gap between having information and having a personal experience of it, for example by providing simulations that can actually improve on the experiences that real life allows. But this must be carefully done and failure to adequately simulate important aspects of the practice

---

[3]Another trend predicted years ago by Mumford, op. cit.; see also Handy, Charles. *The Age of Unreason*, Boston: Harvard Business School Press, 1989.

world is a danger—and a likelihood—given our current understanding of how to apply these technologies. On the other hand, technology has already widened that gap, promoting a population that prefers to experience the world vicariously via the orthicon tube; a world filtered, by and large, to emphasize the sordid and the sensational. And, again, we're on the threshold of that new information infrastructure. The impact of so much information, so much programming seeking to claim our attention and our dollars, so much life-forming experience (however vicarious), will be multiplied by sheer numbers and by the very nature of the interaction. The impact will be enormous.

But I digress.

THREE ELEMENTS OF THE NEW MEDIA: DESIDERATA AND DANGERS

The new media can be viewed as a convergence of—and synergy among— three technologies: computers, electronic media, and communication networks. Here are some of the salutary uses—and dangers—that might be found in these.

Computers

Computer technologies can assist us in gathering, storing, retrieving, displaying, and analyzing data to get some information; perhaps some knowledge. Those data can be more than simply text. For example, x-rays and scans, microscope slides, even recordings of surgical procedures (many of which are already performed in a video environment) could be accessed. New methods of visualizing data, including interactive statistical graphics and scientific visualization techniques, could be applied to help navigate, explore, and analyze complex, multimedia data sets. Further, much of this work could be automated, with intelligent agents—software routines with a persona—acting on behalf of an analyst or policy maker to ferret out salient features of a data set, cases worthy of audit, deriving outcomes-related knowledge to feed back into the care system in an efficient and very timely way. Educational interventions can be planned, even constructed, based on the raw and reduced data and information that have been gathered and analyzed. And these interventions can employ computer-generated interfaces to information and experiences using multiple media. A result can be an effective and efficient, tight feedback loop connecting clinical practice, research into the outcomes of care, and devising outcomes-specific educational interventions that are targeted to the individual care provider to positively affect performance.

On the other hand, there are problems in adequately formulating questions and sampling the world of clinical experience in ways that include important phenomena, particularly subjective ones; in ensuring that those data we do gather are sufficiently complete and accurate, that the analyses are appropriate and well-understood; that the information and its application are well-considered—and tempered with human caring and judgment. There is great danger in not keeping these requirements—and the limitations on being able to satisfy them—firmly in mind.

> In a culture in which the machine, with its impersonal and endlessly repeatable operations, is a controlling metaphor and considered to be an instrument of progress, subjectivity becomes profoundly unacceptable. Diversity, complexity, and ambiguity of human judgment are enemies of technique.
>
> [B]ecause the computer [is treated as if it] "thinks" rather than works, its power to energize mechanistic metaphors is unparalleled... We have devalued the singular human capacity to see things in whole in all their psychic, emotional, and moral dimensions, and we have replaced this with faith in the powers of technical calculation. (Postman, 1992, pp. 118, 158)

The hazards are amplified by the rapidity and efficiency with which such a mechanized system can turn around data, qualities that we generally value and wish to augment. A tendency to automate without due regard to quality and meaning, combined with the tightness of the feedback loop, can result in a system that goes rapidly out of control.

Media

In the world of entertainment, and to a lesser extent in education and training, the craft of creating impactful, engaging presentations has been honed to a very fine edge. Sounds, video, computer animations can make for very memorable experiences. We now have a mature art and method of recording, or creating, human stories that can, in the best cases, convey meaning and insight, even engender wisdom. Combined with sound pedagogy and married to computers, the art and technology of media can provide interactive learning that is memorable, that combines the best that education and media have to offer, to help make complex concepts clear and to bring the complexities of real human experience more immediately to the discussion than usually happens in the typical medical or nursing school lecture hall.

On the other hand, these powerful tools have often been used inappropriately to persuade rather than to educate; subtle or blatant, biased information and propaganda of various shades—commercial, political, or for the public good—are widely used to shape public and private decision making.

Health care, subject to powerful economic forces and vested interests, is no stranger to these kinds of bias. Consumer health information services are likely to become a lucrative market, and network capacity will initially create a programming vacuum that many commercial interests will seek to exploit. And with interactive technologies the presentation can be even more powerful. Great care must be taken to assure that information provided to decision makers, be they clinicians or patients, is as free of content and framing bias as possible. It's unlikely that this will happen without vigilance on the part of public interest agencies.

Use of media tends to require special talent and expertise. A very close collaboration among educators, health care experts, and media experts is required. This need intensifies when large issues of content accuracy, framing, presentation, and balance must be dealt with, as commonly occurs in health education. Unfortunately, this close collaboration is a rare event, usually due to a lack of common experience and vocabulary among the development team, exacerbated by the frequent unavailability and competing priorities of busy clinician content experts. Finally, it is frequently difficult or impossible even to view, much less evaluate, complex, computer-based, multiple media programs; fully exploring numerous pathways that can branch in numerous ways and reviewing numerous optional segments is a labor that few, if any, outside reviewers or students will endure or do well.

Communication Networks

As we've heard, there is an inexorable push to establish high-bandwidth digital networks capable of conveying information existing in multiple media, including motion video and sound. Moreover, these networks will be interactive, so that passage of information is no longer passive and one-way. The possibilities for health care are enormous, particularly for citizens and practitioners in underserved areas (e.g., rural and inner city). Important applications include those that will help ease the administrative burden of the rural practitioner; those that will assure completeness, accuracy, and ready availability of an individual's health record; those that will provide consultative services at a distance to avoid unnecessary travel and hardship to the rural citizen and his family, and to support our rural practitioners; and those that will help the rural practitioner feel part of the larger health care community in our region; and to provide educational services tailored to the performance and needs of the practitioner.

However, there is a fair likelihood that certain populations, such as rural citizens, will not have this service any time soon. The cost of putting up satellites, and their limited bandwidth, combine to require physical links to rural communities. This is likely to be left to the carriers who have few incentives to expend resources for what they consider a limited market. New regulatory environments will obviously play a role, but there is reason to fear that there will be fewer, rather than more, incentives for rural "datafication." Universal access, something we've taken for granted with telephony and electrification, may not be there. Many may only have a dirt road, or no road, leading to the information superhighway. Though a longer-term trend to dispersion of the workforce into rural America may eventually lead to rural services, the short-term outlook is very unsettled. This is very much a topic for discussion and activism.

AND FINALLY. . . EDUCATION

Why Health Education Reform?

Here are two of many reasons to consider dramatic change in the way we train our health care providers. The focus is on medicine, but these concerns are generally applicable to education in the health professions.

As discussed elsewhere in this volume, health care reform is impossible without an expanded role for, and increased numbers of, generalist physicians and alternative primary care providers such as nurse-practitioners. The technology, techniques, and venue of medical care are changing rapidly as our society is changing, and the exercise of reform will undoubtedly accelerate the rates of change. There is concern that the current model of medical education is not able to accommodate, much less assist, these changes (Gastel & Rogers, 1989).

These criticisms and concerns stem in large part from an historic and current domination of medical education by acute care hospitals, which focus on the 5% of patients with the most acute pathology, at the expense of the 95% of patient care that occurs outside the hospital. The current, heavy emphasis on inpatient and subspecialty care in clinical medical education is inappropriate today and likely to be so in the future (Boufford, 1989).

> Just as the health care delivery system in the United States has been dominated by acute care hospitals, so, at least since the 1920s, has organized medical education. . . . Most of clinical medical education and of basic science preparation for it have focused on the 5% of medical practice handling patients with the most acute pathology. The inpatient enterprise has been a relatively well-oiled machine, with an intense sense of urgency, considerable dominance of subspecialists applying their substantial expertise to narrow patient problems for which the student sees the effectiveness of a high-technology intervention.
>
> Meanwhile, the student's exposure to the other 95% of medical practice has been less than satisfactory. The occasional rotation in an ambulatory care setting is often in an overcrowded, high-volume, block-appointment clinic (a

relatively chaotic environment for the first-time visitor) or in a private doctor's office (where student participation typically is rather restricted).

Applying the model used for training in inpatient medicine—block rotations—to the ambulatory setting leaves the student with a marginal role to play in a series of first encounters that involve lengthy interviews, physical examinations, and problem formulations, and for which the student sees no outcome, since he or she leaves the rotation before the patient returns. The alternative is a series of acute walk-in encounters where a focal complaint is quickly treated and the student has no knowledge of the patient's past or future. In this setting, the student has a low sense of mastery and may not perceive the efficacy of medical care. Finally, many students find the ambulatory setting less attractive than the inpatient services because physicians are less dominant, other health professionals play important roles, and patients have considerable autonomy. The primary care physician focusing on transaction-oriented medicine greatly dependent on patient participation certainly can compare unfavorably with the procedure-oriented inpatient specialist whose action appears to be the sole reason for patient improvement.

Shouldn't we shift clinical preparation to better reflect, and prepare the physician for, the 95% instead of the 5% of medical practice? (Boufford, 1989, p. 69)

There is also a *dis*integration of medical teaching and learning, with basic science subjects (e.g., anatomy, physiology, biochemistry) usually taught as isolated facts and with little effort to make them clinically relevant. This is typified by the common practice of separating the learning of basic, factual knowledge from acquiring clinical knowledge and skills, and spending two "preclinical" years to accomplish the former (Barrows, 1989).

'Preclinical,' is, nevertheless, an appropriate term for the first two years, as the student is in a classroom listening to a series of lecturers from isolated disciplines disgorge voluminous information, facts and factlets, that the student knows will have to be disgorged on written examinations that challenge mostly the cognitive process of rote memorization, not the process we would like to see all that valuable information hooked up with [clinically applicable problem-solving and decision-making]. Analysis, synthesis, and application of principles to patient-related issues are rarely assessed. The half-life of information learned through memorization is so short that cramming and review are essential before each examination—yet somehow the examination score is felt to indicate that the student has learned...

Integrated courses do not integrate information in the student's mind, as they still treat each discipline in a separate compartment; the integration is only in the calendar. ... No wonder the student arrives in the clinical years remembering little from the basic sciences, and that the information left in long-term memory is not enmeshed with the clinical reasoning process or recalled in association with patient encounters. (Barrows, 1989, pp. 48, 49)

To address these two problems, health care educational reform must include a restructuring of health education to decrease dependence on hospitals and to increase opportunities to gain experience in providing comprehensive and longitudinal care. And, not incidentally, it should assist in recruiting more generalist practitioners by providing experiences and role models for students that will make the generalist specialties more appealing. In the best of worlds, students would have highly meaningful educational experiences that are available to all on demand, that are engaging and intellectually stimulating, and that teach more than facts. In the best of worlds, the experiences would help the student develop not only knowledge, but the wisdom and the artistry that is seen in the best practitioners.

Tall orders. I believe that these goals can, at least partially, be addressed by applying the new media. But they must be applied well, with artistry and wisdom.

The Need for Reform in the Use of New Media

Sitting in on a discussion among new media technologists—be they in computing, media, communications, or education—can be a heady experience. Visions of global interconnectedness, on-line societies and virtual spaces for work and learning, interactive interface designs for knowledge exploration and navigation, virtual human bodies on which to practice surgical technique, virtual data and virtual reality spaces that can lead to more immediate understanding. These concepts are among many that excite and strain our ability to imagine. There appears to be no dearth of technological ideas about developing and applying the new media.

What we do lack, however, is a solid core of educational ideas about how to use the new media to help people learn. This is not unique to new media, of course, but the application of these technologies —and the large sums of money often spent on doing that—accentuates the issues. I don't presume to offer a comprehensive discussion of these issues, nor to be prescriptive. I do wish to emphasize what I see as the greatest potential strengths of the new media: to provide educational experiences that are optimally suited to the individual; experiences that promote reflective thinking and practice, and the development of professional artistry; experiences that emulate—and deliver on demand—the best that great teachers have to offer. In doing this, I will draw on the work of two educational theorists: John Dewey and Donald Schön.

Experience and Education

At a recent educational technology meeting, a woman who controls a substantial training budget viewed an interactive video program that was heavy on exploration and hypermedia. She was appalled at the lack of pedagogical structure and about the apparent lack of intent toward established instructional goals and objectives.

At another meeting, the dinner speaker gave a talk on technology in education, culminating with the notion that we should do less teaching and allow more learning. The student should be encouraged to learn mainly through exploration and discovery. Provide an interesting collection of information presented in a variety of media, provide a means for exploring the information, and stand back. Other than as a coach or facilitator, the teacher should have little or no direct role in the process.

Both individuals had strong views about what constitutes new media and how they should be used. Both views have some validity. Together, these views guide most new media educational programs now being produced. Both views are limiting.

They are limiting because they are primarily concerned with ideology and not as much on what is accomplished in the mind and spirit of the individual. Yet they are mired on opposite sides of the same ideology. By and large, one group is intent on retaining a traditional, assembly-line, goal-based model of education, whereas the other is intent on avoiding it.

A result of these limiting views is educational new media programs that are *mis*-educative. Many of these programs perpetuate a drill and practice, rote approach to education. Others have students constructing knowledge without providing a sequence of concepts that can lay a foundation and create a framework for understanding. Either type may help a student in special circumstances. Neither is generally useful or effective.

There are other ways of approaching this question, setting aside labels and ideology (one hopes). The question is appropriately framed in terms of teaching and learning, and how new media might help individuals and groups to learn about meaningful things, in ways that matter to them. There are two important aspects to doing this, both derived from the work of John Dewey. The first is the role of multimedia in providing experiences for learning; the second is broadly concerned with using multimedia to promote reflective thinking.

John Dewey is arguably the most influential of American philosophers and perhaps the greatest educational theorist of our time. A hallmark of his thinking is "the organic connection between education and personal experience" (Dewey, 1963, p. 25).

Experiences can have a profound effect on what is learned and how learning occurs. Further, experiences occur regardless of the pedagogical approach taken. With traditional approaches, the experiences can be mis-educative.

> It is a great mistake to suppose, even tacitly, that the traditional classroom is not a place in which pupils had experiences, . . . [but these] were largely of the wrong kind. How many students, for example, were rendered callous to ideas,

and how many lost the impetus to learn because of the way in which learning was experienced by them? How many acquired special skills by means of automatic drill so that their power of judgment and capacity to act intelligently in new situations was limited? How many came to associate the learning process with ennui and boredom? How many found what they did learn so foreign to the situations of life outside the school as to give them no power of control over the latter? (Dewey, 1963, p. 26)

Dewey is equally critical, however, of so-called "progressive education" as it is widely practiced. He has been falsely accused of advocating that educators not be overly concerned with teaching and structure, and that they promote experiential learning by providing experiences that are largely random and informal. On the contrary, he views lack of a teacher's guidance and lack of deliberate structuring of content as "really stupid. For it attempts the impossible, which is always stupid; and it misconceives the conditions of independent thinking" (Dewey, 1926). He counsels against the development of new educational approaches mainly in reaction to the perceived ills of earlier ones.

The general philosophy of the new education may be sound. . . . There is always the danger in a new movement that in rejecting the aims and methods of that which it would supplant, it may develop its principles negatively rather than positively and constructively. . .

Take, for example, the question of organized subject-matter. . . . The problem for progressive education is: What is the place and meaning of subject-matter and or organization *within* experience? How does subject matter function? . . . A philosophy which proceeds on the basis of rejection, of sheer opposition, will neglect these questions. It will tend to suppose that because the old education was based on ready-made organization, therefore it suffices to reject the principle of organization *in toto*, instead of striving to discover what it means and how it is to be attained on the basis of experience. (Dewey, 1963, pp. 20–21)

Providing experiences of high quality is key to success in education. Quality has two aspects: an immediate, apparent one of how accessible, enjoyable, and engaging the experience is, and the less easily determined one of the *effect* the experience has on the learner. Effective education lays a foundation for understanding and builds upon it. The resulting structure supports later consideration of information by the student. There are clear, appropriate and impactful examples, so that the student can become involved with concepts viscerally, as well as intellectually. The student is challenged to inquire, to manipulate and build ideas. "Inquiry-oriented learning theorists from Dewey to Bruner and the modern cognitivists view instruction as a way of encouraging students to become active constructors of knowledge, of knowledge as open and evolving, of academic learning as exciting and vital, and of teaching as a stimulus to curiosity and a model of inquiry. . . (Seal-Wanner, 1988).

This brings us to Dewey's idea of an *experiential continuum* possessed by each individual, defined by his or her unique set of experiences and associated learning—in all aspects of life, not just the school. That continuum uniquely determines what further experiences encourage an individual's intellectual growth. It defines the opportunities for learning for that individual at that moment. At best, an educational experience promotes worthwhile future experiences: "Wholly independent of desire or intent, every experience lives on in further experiences" (Dewey, 1963, p. 90). As Dewey quotes the poet:

> . . . all experience is an arch wherethro'
> Gleams that untraveled world, whose margin fades
> For ever and forever when I move.

If each student is different, and each is changing over time, it is clear that traditional, lock-step approaches to education cannot provide experiences that are well- suited to learning. On the other hand, simply providing access to experiences without providing guidance is also to be avoided. The best course is one taking the individual characteristics of the student into account, while structuring his or her environment and contexts for learning in ways that build positively on past experiences. *It is precisely that course that allows us to escape the limitations outlined at the outset.*

Dewey acknowledges that this is more easily said than done, and that one barrier to adoption of these principles is the difficulty in planning, developing, and implementing methods to apply them.

> The educational system must move one way or the other, either backward to the intellectual and moral standards of a prescientific age or forward to ever greater utilization of scientific method in the development of the possibilities of growing, expanding experience. . .

> The only ground for anticipating failure in taking [the latter] path resides to my mind in the danger that experience and the experimental method will not be adequately conceived. There is no discipline in the world so severe as the discipline of experience subjected to the tests of intelligent development and direction. (Dewey, 1963, pp. 89, 90)

And here lies a major sticking point. We haven't yet established methods for developing sequences of educational experiences that educate and move, that make learning meaningful and memorable. There are, of course, great teachers and they've been around for a very long time; but we don't really know what makes them so effective. And there are theorists, but few who actually apply these theories. Some educators, and some new media developers, appear to have a natural ability to make progress toward Dewey's ideal. They're the exception.

As with education in general, this is *the* key issue in the application of new media in professional education: whether an ability to craft life-forming "experiences subjected to the tests of intelligent development and direction" will remain the province of a few developers and great teachers, or whether we will we be able to devise proven, replicable, and learnable methods more generally to develop educational experiences that meet Dewey's ideal. (An alternative, related question is whether the new media might be applied as a way of channeling the best that great teachers have to offer, but this begs the same issue: how best to do it.)

Nevertheless, with new media it is today possible to construct experiences that are engaging (even gripping), realistic, and comprehensive in dealing with complex issues (Henderson, 1990, 1991). Students can have vicarious experiences of a wide variety, limited primarily by the imagination and budgets of the developers. Moreover, having expended the effort in their construction, these programs are readily available at any time and nearly any place, with the experience known and, ideally, of high quality. If carefully prepared to be accessible to the average teacher and student, these approaches may provide a means to bring Dewey's ideals to the average student in a widespread way. If properly applied, new media could serve as a medium for students and teachers alike to gain access to a world of appropriately structured, vital, growth-enhancing experiences.

Reflective Thinking, Reflective Practice, and Professional Artistry

The artistry of painters, sculptors, musicians, dancers, and designers bears a strong family resemblance to the artistry of extraordinary lawyers, physicians, managers, and teachers. It is no accident that professionals often refer to an "art" of teaching or management and use the term artist to refer to practitioners unusually adept at handling situations of uncertainty, uniqueness, and conflict. (Schön, 1987, p. 15)

Most of education today places great emphasis on memorization of facts and theories, justified to large extent on preparation for standardized tests of knowledge. The recitation (orally or through multiple choices) of abstract, theory-based knowledge is part and parcel of formal education in most of the world today, whether in grade 4, 12, or 18.

This situation goes directly to habits of thinking, learning, and acting and to how these habits limit opportunities for growth and real-world problem solving. By and large, in education we approach the world as modern, scientific thinkers, believing it to be amenable to abstraction—and containing deterministic problems with clear, textbook solutions. There is little, if

any, regard paid to the particular case, nor to the artistry of framing problems, formulating and implementing solutions, estimating and improvising
while dealing with real-world, indeterminate situations.

And students mostly buy into this abstract, deterministic view of the
world and how to deal with it. They're rewarded or penalized according to
their ability to produce answers, either through recall of memorized facts or
through solving problems that almost invariably have clear answers. Some
students are successful and move up the academic ladder; others aren't and
drop out, mentally if not physically. As a result of this recitation approach to
learning and thinking, many—perhaps especially those climbing higher on
the academic ladder—can be ill-prepared for the real world of human
experience. This extends to education for the professions; we don't do a very
good job teaching the art of professional endeavor.

Reflective Practice and Professional Artistry.   Donald Schön wrote
about education from the viewpoint of practice in professions such as
medicine, law, business, engineering, teaching. Schön (1987, p. 3) contrasted
the high ground of "manageable problems [that] lend themselves to solution
through the application of research-based theory and technique" with the
swamp of "messy, confusing problems [that] defy technical solution."

> The irony of this situation is that the problems of the high ground tend to be
> relatively unimportant to individuals or society at large, while in the swamp lie
> the problems of greatest human concern. The practitioner must choose. Shall
> he remain on the high ground where he can solve relatively unimportant
> problems according to prevailing standards of rigor, or shall he descend into
> the swamp of important problems and nonrigorous inquiry?

> The dilemma has two sources: First, the prevailing idea of rigorous profes-
> sional knowledge, based on technical rationality, and second, awareness of
> indeterminate, swampy zones of practice that lie beyond its canons. (Schön,
> 1987, p. 3)

Schön went on to point out that outstanding professional practitioners,
those who deal well with the swamp, aren't generally said to have more
knowledge than others. Instead, they're described as having more wisdom,
talent, intuition, or artistry. But, he said, these are commonly regarded as
junk categories, as phenomena that aren't amenable to scientific tests of
rigor or relevance. As a result, professional education believes it cannot
adequately deal with them. He disagreed:

> On the basis of an underlying and largely unexamined epistemology of prac-
> tice, we distance ourselves from the kinds of performance we need most to
> understand. . .

The question of the relationship between practice competence and professional knowledge needs to be turned upside down. We should start not by asking how to make better use of research-based knowledge but by asking what we can learn from a careful examination of artistry, that is, the competence by which practitioners actually handle indeterminate zones of practice. . . (Schön, 1987, pp. 13–14)

Schön described artistry as a kind of intelligence that is inherently different from, but essential to the exercise of, standard professional knowledge.

In the terrain of professional practice, applied science and research-based technique occupy a critically important though limited territory, bounded on several sides by artistry. There are an art of problem framing, an art of implementation, and an art of improvisation—all necessary to mediate the use in practice of applied science and technique. . . (Schön, 1987, p. 13)

Schön raised and addresses these deeper educational questions: Can any curriculum adequately deal with the "complex, unstable, uncertain, and conflictual worlds of practice?" Can anyone, having studied and described it, teach artistry by any means?

Several educational theorists from Dewey to Schön have emphasized the importance of reflective thinking in education. Dewey (1933, p. 3) described reflective thinking as "the kind of thinking that consists in turning a subject over in the mind and giving it serious and consecutive consideration." He elaborated:

. . . *reflective thinking,* in distinction from other operations to which we apply the name of thought, involves (1) a state of doubt, hesitation, perplexity, mental difficulty, in which thinking originates, and (2) an act of searching, hunting, inquiring, to find material that will resolve the doubt, settle and dispose of the perplexity. (Dewey 1933, p. 12)

Reflective thinking, in other words, is the best approach to indeterminate questions and problems; problems that are complex and messy.

Extending this idea, Schön developed a concept of reflection-in-action, which is distinct from another concept; knowing-in-action. Knowing-in-action applies our existing knowledge to expected situations. Reflection-in-action comes into play when a situation that develops falls outside the boundaries of what we've learned to consider normal. That is, we're surprised by a development in the situation. Surprise leads to reflective thinking as Dewey has outlined; we embark on a process of critical examination of the situation, framing of the problem, and on-the-spot information-gathering (experimentation) to explore the new phenomena. The experimentation may work, yielding expected results, or it may lead to surprises that call forth additional cycles of reflection-in-action. Schön acknowledged that this is idealized; however, he feels that here lies a phenomenon that sums up the elusive artistry discussed previously. Reflective practice is professional practice that uses reflection-in-action to deal with problems "in the swamp." The phenomenon is subject to study. Moreover, Schön proposed that reflective practice can be learned through exercising reflection-in-action, and that professional education can and should provide opportunities for doing so. (Dewey, of course, advocated doing this throughout the course of learning and growth, from the earliest years.)

Reflective Practicums and the New Media.   Schön argued eloquently for the use of "reflective practicums" to promote development of reflection-in-action. He described the reflective practicum as:

> . . . a setting designed for the task of learning a practice. In a context that approximates a practice world, students learn by doing, although their doing usually falls short of real-world work. They learn by undertaking projects that simulate and simplify practice. . . . The practicum is a virtual world, relatively free of the pressures, distractions, and risks of the real one, to which, nevertheless, it refers. . . . It is also a collective world in its own right, with its own mix of materials, tools, languages, and appreciations. It embodies particular ways of seeing, thinking, and doing that tend, over time . . . to assert themselves with increasing authority. (Schön , 1987, p. 37)

New media can be used to create effective reflective practicum experiences. Carefully constructed, replicable, readily available, interactive systems can provide complex and highly involving simulations of real-world situations. An example is a program produced for the Navy, *Regimental Surgeon.*

New Media and Reflective Practicums for Health Education: An Example. As noted previously, much of medical education takes the high ground, and avoids the indeterminate problems of the swamp, particularly the 95% of clinical practice outside the hospital. This is not to say that theory-based knowledge has no value. It clearly does. However, the knowledge must be applied in the real practice world if its relevance is to be seen.

An example "in spades" of this problem, one that highlights these issues (and one that, conveniently, has been addressed using new media) is the plight of the military physician. He is educated in one profession (medicine) and must, on demand, work in another profession (the military) with little or no preparation. This can be an enormously disorienting experience, involving very complex interactions among many factors about which he has little knowledge, and over which he has little control. There are conflicts of values between the respective professions, with military values nearly always prevailing. In this environment the military physician is expected to know his main profession and perform at the level of an expert; in reality he is simply not prepared to deal with operational medical issues that, almost without exception, fall into the swamp category.

Regimental Surgeon (Henderson, 1990) is an interactive multimedia program that strives to provide a reflective practicum experience according to these criteria. The program is intended to foster facility in applying reflection-in-action in the professional activities of the student, and to promote the development of new knowledge and attitudes through an exercise of reflective practice in a multimedia virtual world. It teaches through experience by putting the learner on the job, in this case as regimental surgeon responsible for the health of 3,000+ men in a combat zone. The learner exercises *knowing-in-action* in more routine medical matters, but is frequently challenged by a series of surprises designed to throw him into *reflection-in-action*. The user can enter, explore, discover facts, and formulate rules and principals; this, much as one would in real life. The program is intentionally entertaining as well, placing the learner within the context of a story whose evolution and outcome depend on the decisions made by the learner. Numerous themes are developed simultaneously, any of which can be developed and emphasized in discussions outside the context of the program itself.

The program has the following objective: given a hypothetical combat scenario involving U. S. Marine Forces, to provide, through simulation, experiences which will allow the learner to develop knowledge and skills needed to function as a staff medical officer. In the process, the learner is challenged to: (a) evaluate and define the medical threat (malaria) confronting Marine Forces in his unit; (b) identify, plan, and implement the preventive medicine countermeasures which can reduce risk; (c) communicate the urgency of those countermeasures to key commanders.

There are 10 other facilitating objectives dealing with basic organizational, interpersonal, and medical aspects of functioning as a staff medical officer.

The program is organized somewhat as an adventure computer game, where various locations can be visited. In such a game, the locations can contain treasures and/or characters with which limited interactions can occur. In *Regimental Surgeon*, the locations are various elements and units of the regiment (headquarters and the three battalions), division headquarters, and medical support units (hospitals). The "treasures" are facts that can be obtained from a variety of sources in a variety of ways: epidemiologic surveys, reading documents, questioning individuals, looking at blood smears. The learner is required to assemble the facts he's acquired, draw conclusions, and, based on these, inform and make recommendations. Characters can facilitate or obstruct this process, or act as mentors, depending on the learner's actions as he traverses the program.

The program begins with an opening sequence which sets the scenario, introduces most of the main characters, and establishes principal rules of the game. It is also intended to engage the learner and motivate him to continue and complete the simulation.

Midgame contains the body of the program. There are 40 scenes of two types: initial and developmental. Initial scenes are those seen by the learner on his first visit, whereas developmental scenes further plot development and reflect learner's progress to that point. During development there are up to three encounters with each of eight main characters (of a total of 32 speaking parts). For seven of these characters there are different versions of the encounters, encouraging/friendly or distracted/distant or unfriendly, depending on how the learner has handled his job so far. Each character, representing main teaching points, has a different view on the learner's job and how he should be doing it. Response of the other principal character (the chaplain) is constant for each of the encounters, though his dialogue depends on the learner's previous choices.

Epilogue is a coda to wrap things up. If the learner has managed to keep Division out of things and has delivered his briefing, he gets a chance to back up his assertions with evidence; feedback depends on completeness of investigation and plan. If Division did get involved, then a different ending is seen.

Because there are many ways to traverse the program, the learner can gather information in ways and to an extent that is variable (the role of chance is minimized by placing critical information redundantly). For example, the learner may choose to play a very active role, traveling personally throughout the command, gathering needed information and completing the program in a very comprehensive (perhaps too comprehensive) and successful way; or he may delegate some of the work and play a less active role, and still be successful; or, he could choose to stay in his tent, or sleep, with an unsuccessful outcome. In each case the learner experiences the consequences of his actions, and learns whatever he decides is the knowledge contained therein. Thus instructional content is variable in two ways: it depends on the path chosen by the student and it depends on his interpretation of the events encountered on the path.

Finally, representing contained information is an interface design issue addressed by what some call a *concept map*. Concept maps in *Regimental Surgeon* are literally a series of nine animated maps which provide a means of navigating through the physical, temporal, and concept elements of this reality. Upon reaching a location, the user is presented a "talk" menu allowing him to ask questions or read documents, blood smears, or tactical maps. A design goal that the interface be intuitive, that is, its use not requiring additional instruction, was met in two ways: (a) by using logical design, in the sense that graphics and content are consistent and integrated with the reality portrayed; and (b) by incrementally increasing the complexity of the interface, starting with simple menus and adding features.

Unfortunately, we don't know if these educational goals have been realized, because there has been no study of the program in this regard. (This gets into another tangle of issues regarding evaluation that we will side-step here, thus perpetuating a very unfortunate practice to neglect this important issue.) The program has been widely disseminated and there are anecdotal reports that it was helpful in preparing some medical officers for duty in the Persian Gulf War. It is an early example of the use of new media and virtual environments to promote reflective practice and the art of dealing with the swamp of professional endeavors.

Another Example: HIV/AIDS

Our laboratory has just completed a project titled *HIV & AIDS: An Interactive Curriculum for Students in the Health Professions.* This interactive multimedia program combines theory-based education concerned with factual knowledge, with a series of reflective practicums consisting of simulated patient encounters. The goal is to cover all major aspects of HIV and AIDS, from the molecular level to the complex psychosocial, integrating the presentation of factual information with clinical application.

HIV & AIDS pays great attention to esthetics and the quality of media production and programming. Graphics interfaces are easy to use and responsive, 2- and 3-D animated graphics are used to help make complex concepts, such as HIV binding to $CD4^+$ T-cells, clear and memorable. There are very moving—even riveting—interviews with real individuals; four with HIV and six care providers. Of the former group, two are married, one is a gay nurse, and one is a mother who lost her child to HIV. The program has approximately two hours of full-motion video and an additional two hours of audio to support voice-over text and graphics. There are numerous exercises and activities in which the student interacts with video, text, and graphics; the goal is to have students participate actively as they progress through the program.

The reflective practicum consists of a series of four encounters with a simulated patient, Laurie Matthews. Ms. Matthews is a young woman whom we manage from her initial, pretest counseling visit, over a period of four years, to a final visit when she is diagnosed as having AIDS and is admitted for pneumocystis carinii pneumonia. Learners are told that the patient is an actress. However, it is remarkable that most become very emotionally involved in her story, which unfolds in a series of disclosures; these, to some extent, are governed by the choices of the learner.

Each encounter consists of gathering clinical information from the patient, followed by a series of questions posed by Ms. Matthews to the learner. Interaction is via menus with varying numbers of choices. For each of these questions, the learner can obtain feedback and discussion of issues; this "mentor" function can be activated by the learner as desired. Though some questions require mostly factual explanations, all are concerned with the art of patient counseling. Most involve complex issues in the swampy world of HIV practice: fears of death and rejection, guilt, relationships and marriage, pregnancy, noncompliance in medication and safer sex practices, ethical issues for the practitioner (framing, paternalism, patient autonomy).

New Media for Great Teaching?

There are, of course, data and information bases from which to learn. These will proliferate, become richer in content and media used, and become—electronically and conceptually—more accessible. They will play an increasingly important role in health education. However, it is important to distinguish between these systems and new media programs that are specifically designed to be educational, that do more than simply provide the information. In these programs, experiences are crafted—in the Deweyan sense—specifically to help people learn; in other words, they teach.

This view of a new media program that teaches is somewhat out of favor. Many prefer the database, explore-and-discover model, as discussed previously. But this philosophy avoids or ignores the challenge and flat-out excitement that a great teacher and a wonderful sequence of ideas can bring to a child or adult who must learn about something.

The best teaching programs can accomplish much of what a great teacher can do. They can excite interest; explain clearly with great, even stunning examples; allow learners to manipulate, play with, and develop ownership of concepts; and provoke reflective consideration of them. For a majority of students, these don't happen simply by providing access to a large, hypermedia data/information set, no more than providing access to a library or museum.

This kind of new media program, to be complete, should have two more things. First, they must have what Dewey called "a career"; that is, they should give a strong sense of participating in a reality that has a past and present, and is going somewhere. In other words, it exists in time and space, if only in our imaginations. Second, the program should in some way enter and occupy a place in our hearts; there should be an emotional richness that speaks to our human caring. These can provide a unifying framework for a program's concepts and, at the same time, provide information that can't be codified, quantified, or computerized. That is, the human, rich, swampy, nondeterministic stories of people and their situations can reveal the consequences of decisions made and actions taken, not simply their process.

There's nothing magical about doing this; we've been telling stories to pass on information, knowledge, and even wisdom for centuries. As we bring to bear increasingly sophisticated and complex technologies, we've been doing it with increasing levels of sophistication, complexity and—for good or ill—effectiveness. The danger lies in not exercising sufficient care in the application of these technologies, resulting in further fragmentation and sterilization of information, and isolation of the practitioner from those he or she would help. The promise is in providing educational experiences that transform our understanding of our world, of ourselves and others, and of the consequences of our actions. There's some wisdom in that.

REFERENCES

Barrows, H. S. (1989). (Commenting on Wilson, M. P.) The clinical teacher and the learning process: Toward a new paradigm for clinical education. In Gastel, B. & Rogers, D. E. (Eds.), *Clinical Education and the Doctor of Tomorrow* (pp. 47–52). New York: The New York Academy of Medicine.

Boufford, J. I. (1989). Changing paths and places for training tomorrow's generalists. In B. Gastel & D. E. Rogers (Eds.), *Clinical education and the doctor of tomorrow* (pp. 67–80). New York: The New York Academy of Medicine.

Dewey, J. (1926). Individuality and experience. *Journal of the Barnes Foundation, 2,* 4.

Dewey, J. (1933). *How we think: A restatement of the relation of reflective thinking to the educative process.* New York: Heath.

Dewey, J. (1963). *Experience and education.* New York: Macmillan. (Original work published 1938)

Gastel, B., & Rogers, D. E. (Eds.). (1989). *Clinical education and the doctor of tomorrow.* New York: The New York Academy of Medicine.

Handy, C. (1989). *The age of unreason.* Boston: Harvard Business School Press.

Henderson, J. V. (1990). Designing realities: Interactive media, virtual realities, and cyberspace. *Multimedia Review,* 47–51.

Henderson, J. V. (1991). Virtual realities as instructional technology, *Journal of Interactive Instructional Development,* 24–30.

Mumford, L. (1934). *Technics and civilization.* New York: Harcourt, Brace & Co.

Postman, N. (1992). *Technopoly.* New York: Knopf.

Schön, D. A. (1987). *Educating the reflective practitioner: Toward a new design for teaching and learning in the professions.* San Francisco: Jossey-Bass.

Seal-Wanner, C. (1988). Interactive video systems: Their promise and their potential. In R. O. McClintock (Ed.), *Computing and education: The second frontier.* (p. 23). New York: Teachers College Press.

V
POTHOLES ALONG THE
INFORMATION SUPERHIGHWAY

But Will the New Health Media Be Forthcoming?

Francis Dummer Fisher[1]
The University of Texas at Austin

The other chapters in this book detail possibilities of how new electronic information media might facilitate improvement in American health care. This chapter cautions that we cannot take for granted the technology needed to realize those prospects. In particular, it is questioned whether free market forces alone will produce either the national information infrastructure (NII) with the power and reach needed to serve health purposes or the health information products which all Americans should be able to access by means of such media. The chapter concludes that to achieve the needed powerful universal network and information products, it is not enough to rely on competition in the free market; strong public leadership and government action is required.

The first part of the chapter deals with the media necessary for health use, spelling out the functionalities required of the system and the need for a universal reach to serve all Americans.

The second part of the chapter deals with the needed health information products, and examines whether prospective health reform efforts adequately assure their development.

COMPETITION ALONE MAY NOT BRING THE NII TO EVERY AMERICAN

Some uses of the new electronic information technology in aid of health do not require a universal network. Computers, networked within an institution, for instance, permit centralized but shared record keeping. Self-standing computer-based instructional units can help train medical personnel by

---

[1]The help of Steven J. Downs in the preparation of this chapter is gratefully acknowledged.

simulating patient pathology. Local area networks improve staff consultation and cooperation.

And when networks are extended beyond an individual health institution, the functionality provided is likely a connection to another institution. Links to the national library of medicine permit many doctors to access the comprehensive record of medical literature. Through video teleconferencing connections, doctors can examine patients in remote clinics and, as we hear so often, high speed networks between institutions can transmit x-rays and other medical images for review by distant experts.

The media required for linking medical institutions to effect such purposes will likely come into being. Connecting hospitals and clinics with a broadband network is an announced goal of the national administration. Many institutions, moreover, lie within central urban areas where information traffic is sufficiently dense to attract investors in providing the needed broad-band carriage functionalities.

But connecting isolated care providers and reaching the general public in their homes to assist self-care and to assure the access of information that can maintain health and prevent disease places demands on the character of the infrastructure different from those needed to serve medical institutions alone. Most importantly, the infrastructure must have not only the power to transmit the needed information. It must reach everyone.

An infrastructure with such a universal reach can not be justified for health purposes alone. Just as educators sometimes propose an education network for distance learning, and state government officials a state network to link up central and regional offices, we sometimes hear those concerned with health speak of a health network for health purposes. But education, state government, and health users, when they wish to reach everywhere with a universal service, will inevitably be drawn by economic forces to share a common public network infrastructure. Pieces of the infrastructure, of course, might belong to different entities, but the whole network of networks will be interoperable and universal and used in common by users for different purposes. It is for this reason that those of us concerned with health must focus on the emerging National Information Infrastructure (NII). We must ask whether the NII is being designed and implemented in ways that serve the needs of health as well as the needs of all its other users.

In the section that follows we examine: (a) the particular functionalities of the NII that must be provided to serve health adequately, and (b) the reach of the NII in terms of both geography and the economic status of users.

Throughout this examination we consider whether the needed infrastructure functionalities and information programs are likely to come into being through market forces alone or whether stronger public leadership and intervention are required.

Needed Powers of the Information Infrastructure

Emerging Consensus: An Open, Switched, Interactive Broadband Network, Reaching Every Home

Just as different potential users of the infrastructure are increasingly recognizing the need to share a common utility, so a consensus is emerging as to what that infrastructure should provide: an open, switched, interactive broadband network, reaching every home. These functionalities are now urged by states, computer company executives, cable television, and telephone companies as well as by Canada. (Computer Systems Policy Project, 1993; Ministry of Supply and Services Canada, 1992; NY Telecommunications Exchange, 1993; Smith, 1993; Time Warner, 1993).

The prestigious national research council, the government-supported organization of our leading scientists, recently expressed the needed powers in this way: "A national information infrastructure should be capable of carrying information services of all kinds, from suppliers of all kinds, to customers of all kinds, across network service providers of all kinds, in a seamless accessible fashion" (National Research Council, 1994). The Information Infrastructure Task Force (1993) of the federal government adopted this goal. But subsequently, in an important speech on January 11,1994 of Vice President Al Gore and in related administration legislative proposals, the goal of the administration was narrowed. Its revised goal for broadband switched service is to reach only school classrooms, libraries, clinics and hospitals by the year 2000. (Gore, 1994). For the administration, reaching homes with switched interactive broad-band service is no longer a goal to be sought at any time in the future. This retreat could seriously set back realizing the new communications technology needed for the health purposes discussed in other chapters. That risk becomes clearer as the powers and reach of the NII needed for health uses are specified more particularly.

Specific Powers of the NII Necessary for Health Use

Broadband Capacity. The telephone network and most existing data networks provide enough capacity to the end user to allow for access to information in speech, text, and rudimentary graphics. Upgrading these networks to broadband capacity would permit access to medical information stored in high-resolution image format such as x-rays, EKGs, and photographs of cell pathologies. It would, as discussed in Preston's chapter on telemedicine, permit physicians to observe and diagnose patients at a distance.

An equally compelling need for broadband capacity is to deliver basic health information to persons who find a video "show and tell" format more comprehensible than abstracted texts and numbers. A picture of how the skin reacts to the bite of a deer tick, if it carried Lyme disease, can be grasped more easily than the text definition of a "red papule of median diameter of up to 15 centimeters." By looking at images of benign, cancerous, and dubious moles, someone with a newly noticed growth can tell whether a trip to a doctor is in order.

As discussed by Wennberg, video-based health information can also include the testimony of peers, as in discussing their experiences with the "watchful waiting" choice for addressing benign prostatic hyperplasia (BPH). A high school student, pregnant for the first time, wants to see and hear how others with whom she can identify succeeded in bringing to term a first baby with an adequate birth weight. Media for health use must include video.

Switched, Open, and Interactive.   Whether the message is voice, data or image, the user must be able to choose the content. The broadcast model, in which information is sent from a limited number of sources to all of the public at once, with little choice and no individual user interaction, will not likely provide the specific information that is of concern to a user at a particular moment. In contrast, the telephone, though limited to voice, can connect a user to any one of millions of information sources. What is needed then in order to provide Americans with specific health information on demand is a system of *switched* video.

To be sure, one-way television, at some time over a period of several years, might broadcast a program on the medical problem of concern to a user. But the user, concerned with tick, mole, prostate, or pregnancy, needs to be switched to information now.

Switching also supports an open system, in which anyone with information to offer can mount an information program on the network and make it available for any user to retrieve. Switching maximizes the desirable competition among information service providers and encourages a diversity of information producers. An open, switched system enables an information marketplace that resembles publishing more than it does television: There will be no scarcity of channels (channels, as a concept, may become meaningless) and anyone can "publish" information on the network the way anyone today can publish a magazine or newsletter.

Interactivity grants the user the power to control the information, to select and recursively define in greater specificity what is wanted. In short, it is the power to order around your smart television. Furthermore, interactivity enables the personalization of information. In the example of a self-care or health tutorial, the program can tailor the information provided to meet individual needs by permitting the user to enter personal characteristics, such as age, weight, and sex.

Broadband networks already reach most homes in the form of television. and the economics of entertainment and home shopping are driving both telephone companies and cable television operators (and joint ventures of the two) to enhance their networks with high-capacity optical fiber. For many Americans, a future of 24-hour access to *Home Alone II* and their favorite "M*A*S*H*" episodes looms on the horizon. But whether such broad-band networks evolve into the NII and have the functionalities of switching and interaction needed for social purposes, like health, is problematic.

One risk is that vertically integrated communications companies, which supply both the entertainment programming as well as the networks that carry the programs could have an understandable incentive to play down switching. This is not because broad-band switches are not in prospect. They are. But switching would permit users to choose the content of some other supplier.

If telecommunications companies were only in the business of carrying the messages of others, they might seek the information traffic of all. But for companies providing both infrastructure and information content, providing common carriage does not seem as glamorous or as financially attractive as making movies or inventing ways to substitute video-order for mail-order purchasing. And as the incentive to push their own messages increases, the incentive to carry the messages of competing information providers decreases.

Two developments hold out hope for the open system: the emerging support for a right of interconnection and the fact that a large cable television consortium, Time Warner, Inc., has stated it will adopt an open system for its proposed broadband "full service network" (Time Warner, Inc. 1993).[2]

The "right of interconnection" was first claimed by companies other than the dominant local telephone company, which wanted to provide alternative voice and data services in a way that used parts of the dominant company's network. Interconnection, it was asserted, would increase competition for content and other functionalities of the network.

Interconnection to the information infrastructure is beginning to be seen more broadly. For example, if customers in a region had a choice of two broad-band network providers, an information service provider would want to ensure that all customers had access to its programs, regardless of which network provider the customers chose. Such assurance of customer access to information, regardless of local network, would be important for public information services, including health information.

So interconnection can both open up a vertically integrated network to alternate information suppliers as well as assure users access to all information even when a user may have selected to be connected to only one of several possible networks.

The free market alone will not assure interconnectivity. Hence, the extent

---

[2]During the negotiations looking toward merger, Bell Atlantic and TCI also pledged switching as a way to assure openness. (Smith, R. W., Bell Atlantic/TCI, 1993)

to which interactive broadband networks will support health and other social uses will much depend on the details of interconnectivity rights and on the certainty that public authority will assure those rights. The pricing of interconnection is also critical, for the right to interconnect can be rendered meaningless if a carrier sets unreasonably high rates.

Although the statement of Time Warner in support of openness is encouraging, it will take public authority to assure that carriers do, in fact, offer interconnection and that they charge other content suppliers no more than they charge themselves for delivering their own content. Unfortunately, the administration does not seem to understand that interconnection, just as much as common carriage, requires this regulatory oversight. Provisions in an administration-drafted proposed Article VII of the Federal Communication Act suggested that price oversight of interconnection will not be necessary except for a carrier that dominated the market. But competition of vertically integrated companies provides no assurance of openness to carriage; regulation assuring interconnection will be required (Fisher, 1994).

A technical but important point in the plans of Time Warner is its proposed use of Asynchronous Transfer Mode (ATM) as the switching platform for its new broad-band interactive network. ATM is an emerging technique that permits messages of voice, data and images to be divided up into small pieces of similar digital cells or packets, thus improving the power of the network to handle the different kinds of information. The use of ATM, in the form being developed as an international standard, can facilitate open participation in Time Warner's network, whereas the adoption of a proprietary platform would raise barriers to participation.

Another important aspect of the adoption by cable companies of openness, if effected, is that it moots a difficult first amendment problem. In connection with other issues, cable companies have sometimes claimed that the first amendment permits them, and no one else, to decide what is carried on their wires, asserting that compelled speech is the reverse side of censorship. (National Cable Television Association, Inc., 1993). In an open or switched system, where channels are not scarce, the network provider does not decide which programs to include or exclude—there is room for all.

Upstream Video. Should the infrastructure include the functionality of upstream video, the capacity to send moving images in real-time from any home, as some public interest groups have urged (Hadden, 1994)? This capability would permit the health condition of anyone, wherever they may be, to be visually monitored at a distance. It would also permit anyone to be not only a consumer but a publisher of video programming, including health information and services, as the Electronic Frontier Foundation (1993) and others have urged. A third application is the "videophone," which would support live two-way video communication between individuals. This would facilitate forming virtual teams of separated health professionals to address together a medical problem that all could view.

The new media applications described in this book do not, by and large, require upstream video. The most compelling health application of upstream video would seem to be the monitoring of patients in the home. For this application, having an infrastructure that is capable of being enhanced to supply upstream video when it is needed, such as during a recuperation period or in the terminal stages of an illness, is certainly desirable. That daily access of all citizens to video cameras and upstream broadband capacity holds much potential for addressing the cost, quality, and access issues of health care reform is still speculation.

Other Upstream Capacity. Although upstream video capacity offers the potential for patient examination and for video communication among individuals, other less expensive forms of more limited upstream capacity can be used to facilitate communication among individuals. In particular, only a small amount of bandwidth would be needed to allow individuals in their homes to communicate with their peers via electronic bulletin board systems. Bulletin boards permit the collection and sharing of opinions on various discussion topics and are commonly used by a variety of health-related self-help groups. The Comprehensive Health Enhancement Support System (CHESS) project at the University of Wisconsin suggests the value of peer-to-peer information exchange and support. The designers of the system found that AIDS patients whom they had connected to an inquiry-response system wanted not just medical information concerning their conditions—they wanted to "talk" with each other. In fact, the peer-to-peer discussion feature was far and away the most commonly used element of the system (Pingree et al., 1993). The value of this human-to-human contact is underscored by Vickery, who, in his chapter notes that the greatest gain associated with the new media and self-care could come from combining the new media with the human interaction that is so valuable in self-help groups and decision support services.

Search Tools: Directories and Gateways. Even if users were provided an adequate network infrastructure and plentiful, useful health information, they must still be able to find the information that they seek. Given the correlation between low educational levels and poor health (U.S. Department of Health and Human Services, 1991), those most in need of information are likely to need the most help in finding it.

For voice information, the telephone book works well to find a number for a known name. Telephone directories work much less well for finding information based on its subject. The Yellow Pages work for numbers of easily classified businesses like "taxi" and "plumber", but fail for someone trying to find an answer to a problem. So finding the good health information that is available on the telephone network is difficult. A newly pregnant high school girl will not typically find much under "baby," "pregnant" or "diet for expectant mothers." Only a very few of the more than one hundred "1-800" free health information services listed in a pamphlet published by the U.S. Public Health Service appear in local telephone directories.

A telephone "directory to health services" is offered now by commercial firms in some newspapers and in a special section of many Yellow Pages. But these usually turn out to be listings of only a selected number of businesses which have paid for getting an edge over their competition. No designation makes clear which listed health services are public in nature, and which are a form of advertising. Worse, the commercial directory is often misleadingly designed to look like a public service. A distressed youngster dialing "teenage suicide" may get only a general canned message on suicide and not be told the number for the community-supported suicide hotline that could really provide help.

Finding health information in cyberspace is even more challenging. The Internet is known for the poor quality of its directories, indicating that whereas there may be a collective interest in better directories, individual information providers and network operators may lack the incentives to improve them.[3]

As the information infrastructure evolves from voice and broadcast television to interactive multimedia, and as information sources multiply, locating information may become yet more difficult. Different addresses (i.e. numbers) for the same information now depend on whether the medium is computer, telephone, or television. Electronic technology could help; cross-referencing should be easier than in a printed directory. But as the firms that provide infrastructure increasingly want to sell their own content, it will probably take public action to assure directories that maximize choice.

The very notion of a single initial directory, moreover, may be contested by commercial directory providers although none of them may be willing to offer comprehensive coverage. To overcome such resistance to a shared or universal directory, the FCC, in its "video dialtone" regulations, requires telephone companies providing video services to put in place a common "first tier" directory, a place where all information suppliers can be listed. In the absence of such regulation, the company providing the video platform would have an incentive to skew directories so as to point users to its own

---

[3]The Internet has taken a step forward in upgrading guidance with the creation of the "InterNIC," a family of Internet directory services, in 1993. Note, however, that the InterNIC was established through cooperative agreements funded by the National Science Foundation, leaving open the question of who will sponsor the directory services of the Internet when it is weaned from the public teat.

affiliated information services, rather than to present equitably the numbers for all. It is particularly important for seekers of health information and services to have one starting point which will list all providers, distinguishing public from commercial providers. And, since information easily transits distance, directories of health information sources should be national in coverage.

The Needed Reach of the NII: A Universal Service to Connect Every American

An information infrastructure gains value as it becomes more widely available. In part, this is because as more persons and institutions are connected, any particular user can reach a larger number of others. In part, a service that reaches everyone permits society to accomplish social programs that require the widest distribution of information. If health information, whether for therapy or prevention, is available over a universal infrastructure, not only can individuals get the health information and services they need, but society can gain the advantages of more efficient medical care, reduced disease and a healthier work force.

Most important, universal reach of information advances the public goal of equity; no one is excluded and we avoid a division between information "have's" and "have not's." Equitable access to health information is especially important when health is at stake.

We examine here the potential reach of the information infrastructure, both in terms of geography and in terms of the economic status of users. Note how the effective use of the infrastructure for health-related uses will depend on the solution by public regulators of many of the issues surrounding universal service.

Geographic Reach

In considering the reach of the information infrastructure in geographical terms , it is helpful to divide America into three markets: central business districts, the home market, and rural communities.

Central Business Districts. In concentrated downtown business districts, the compact demand for advanced telecommunications services will be so large as to assure good service. The functionalities that can be provided by the NII are likely to be so valuable to businesses located there that the cost of providing infrastructure in those areas will appear small by comparison. In the downtown market, therefore, we can anticipate multiple networks providing the needed functionalities and competition among them. Already, former monopolies of voice telephone service in some central business markets are subject to competition from multiple "alternative providers". Major health institutions located within the boundaries of these markets will gain from these competitive offerings.

But as noted above where the importance of interconnectivity is discussed, the reach of needed network functionalities, if provided by vertically integrated companies, will not by itself assure openness to all information.

Home Market.   In the home market, that is, in all the rest of the "approximately 95% of American households" to which subscription television is now available (Turner Broadcasting System, 1993), in essence everywhere except the central business districts and some rural areas, the investment in a more dispersed network is larger in relation to the value of the traffic. Although it is beginning to appear that the band-width of a network in the home market will be raised to the level provided by fiber optic cable, in order to increase the number of entertainment and home-shopping channels deliverable, it seems unlikely that the home traffic will justify more than one such line. The struggle to see which enterprise provides the network serving the home market, then, will not likely result in an ongoing competition over the provision of the functionalities needed in the NII. Particularly problematic are the functionalities of two-way communication and switching.

What seems clear is that the vigorous struggles now observed between those wishing to serve the central business district and those who wish to become the dominant provider of the broadband connection to the home do not relieve public authority from the duty to ensure interactivity and switching, and to protect the right of interconnection and its nondiscriminatory price. Without clearer recognition of this regulatory duty by government authorities, the reach of the functionalities of the NII needed for health uses, particularly interactivity and switching in the home market and openness in both home and central business markets, is not assured.

The Rural Community.   Although fiber optic cable together with either a coaxial cable or wireless "drop" to make the final connection to a home terminal may be the technology of choice for the home market, the technological solution for providing the needed interactive broadband functionalities to rural homes now served not by wires but by one-way satellite or other broadcast television or which have no television service at all seems less certain.[4] The sparse concentration of homes in rural areas raises the cost of network connections substantially.

An interactive broadband connection to some one point in most rural

---

[4]It is the undesignated number of rural homes that are reached with subscription television by satellite or other means of broadcast, rather than wire, that raises the number of homes accessible to "cable" TV to 95% of all homes.

communities may evolve relatively easily. This connection could be shared by those rural facilities concentrated at that point, most likely a school and some sort of health facility. (Office of Technology Assessment, 1991). It was noted, in the study of the Texas Telemedicine project, that in rural Giddings only a modest broadband connection to the clinic from a centrally available communications node was required. How to provide the interactive switched video connection to widely distributed rural homes, however, is not obvious and needs urgent study.

Economic Reach

Affordability: Subsidies To Assure Universality.   Regardless of the geographical area in which they live, some Americans will not be able to pay the market price of the information services society regards as basic. Yet reaching these persons is desirable both as a matter of social equity and to maximize the public benefits of everyone receiving certain information and services, a value especially evident in the case of health information.

Local telephone service has long been provided by a regulated monopoly, allowing the regulatory authority both to demand universal availability and to fix rates to assure affordability.[5] However, as noted, the interactive broadband networks of the future will in some markets be provided by competing carriers, thus requiring a system under which universal service obligations could be shared among competitors, both by their contributing to the universal service subsidy fund and by helping deliver subsidized service. To arrange the funding and delivery of universal service with so many firms providing pieces of the infrastructure will require effective public oversight, another instance where the free market alone must be supplemented by government regulation.

The most difficult policy question related to universal service is the determination of the services that should be made available and affordable to all. Until recently, universal telephone service has been a relatively straightforward matter—in the main, one either had it or one did not. Today, the range of services that may be available grows broader and broader as data and video are mixed into the options. Should the determination of universal service be technical, as in assuring a certain amount of bandwidth (1 Mb/s?, 10 Mb/s?, 100 Mb/s?), should it be functional, assuring two-way video capability (but at what resolution?), or should it be based on

---

[5]Providers of traditional cable television service have not generally been subject to any universal service obligations under federal law. But municipal franchises, under which cable service is provided, often require service to all persons within the franchise area who ask for it. Providing social subsidies to assure that all, regardless of means, can receive broad-band services would represent an extension of universal service, although, in fact, few persons seem unable to finance a TV terminal themselves.

content, guaranteeing access to information of public concern, such as health information (but in which media: text? video?)? As discussed earlier, access to health information, a universal service that provides switched access from the home to video is desirable. But the time by which that goal can be reached needs to be scheduled by public authority. And the definition of what services should be universal will need to be reviewed on a continuing basis as new services and applications come into being. It is discouraging that as of mid-1994, the national administration has not endorsed interactive switched video service for every American even as a distant goal. Advocating universal health care but not supporting the universal information infrastructure that could help achieve it seems inconsistent.

HEALTH INFORMATION AND HEALTH CARE REFORM

The provision of an open information infrastructure reaching everyone with all the desired functionalities will not alone be sufficient to assure its use for health purposes unless health information and services are developed. This concern can be paraphrased as "what if you build it, and they come, but there's nothing there?" What good is an information superhighway without any information? Will the development of what we might loosely call the "content" of health information be sufficient if left to private enterprises? Or will there be a necessary role for government in the development and maintenance of health-related information? And how do the proposals for health reform affect the incentives and responsibilities for developing and distributing health information?

As we examine these questions, three kinds of health-related information can usefully be distinguished: medical records and the "paper-work" of health care; medical knowledge important to professional care-givers; health information for the general public.

In this section, the focus is primarily on the third category. It should be noted, however, that the distinctions between categories are not meant to be rigid, for much of the information itself tends to overlap. For example, information for the public about medical treatment options or disease prevention strategies is derived from the more technical information directed toward physicians and biomedical researchers. Furthermore, as both Vickery (chapter 3) and Wennberg (chapter 6) discuss, an individual's health record is an integral part of the health information that supports his or her decision-making process.

The development of computerized medical records and electronic transaction systems will likely be guided by a combination of business motives and government action. To the extent that computerized medical records are shown to be cost-effective, privately run health care providers would have the incentive to adopt them.[6] The federal government is attempting to encourage development of these systems by providing research and development funds (through the agency for health care policy and research and the national library of medicine) and by expanding electronic reimbursement for medicare to cover some electronic information expenses.

The dissemination of medical information to health care providers has long been primarily the mission of the national library of medicine (NLM). One would expect that the library will continue its role in a reformed health care system and that it will use the NII to open its holdings to greater access. The challenge for the NLM is to (a) exploit the new media and the NII to improve the quality of the information it provides and (b) to facilitate access to that information without leaving certain populations behind. Exploiting technology and at the same time assuring access may be difficult because the opportunities created by the new media and by a richer NII create the potential for widening the gaps between the high-end and the low-end users. These wider gaps may require that the information be stored in "tiers" keyed to the bandwidth and processing power resources available to different groups of users until broad bandwidth power is universally distributed.

Health Information for the Public

As illustrated in this book, health information to the public at large can augment the power of health-providing organizations and professionals, show individuals how to take care of their own health needs, help maintain healthy lifestyles and prevent illness and accidents. Under a reformed health care system, who will produce and maintain these programs of health information for general use—private health organizations, public health agencies, or both?

Provision of Health Information by Private Health Providers

For health providers, using the information infrastructure to inform patients would seem to be a natural extension of providing health information in a clinical setting. Just as a doctor might telephone to check the progress of a patient and to offer advice, more comprehensive and intelli-

---

[6]As Zallen points out in his chapter, one cannot discuss the idea of computerized, networked patient records without raising the issue of patient confidentiality. Zallen points to the success of the Inter Practice Systems pilot and notes that automated systems can be made secure. Ensuring that they will be secure, given the ease of transporting digital records over a widespread information infrastructure, requires thoughtful public policy that balances the advantages of sharing information among providers in support of a patient with the risks of allowing sensitive information to fall into the wrong hands.

gent programs of health information can find out a patient's condition and return appropriate recommendations for care.

An example of such an approach is illustrated by Barry Zallen (chapter 2) on the Harvard Community Health Plan's provision of networked automated health information to its subscribers. Carefully constructed protocols lead a customer to specify symptoms. Then, through interaction with the information program, many of a patient's problems can be more narrowly defined, leading to specific suggestions concerning care.

As reported by Dr. Zallen, patients can, at any time, initiate a connection to a live technician. And, in some cases, the customers' answers to packaged questions would force a human connection—as with a positive response to the question "any chest pains?" Fruitless trips to the clinic can be avoided, as when a customer receives formal advice that a visit cannot improve a diagnosis if a cough has just begun and is accompanied by no pain or fever— "Check with us in 12 hours".[7]

The protocols used by this health maintenance organization were developed in cooperation with the staff doctors. Advantages of having the content of health information be the responsibility of the caregiver include clear accountability and ongoing maintenance of quality and currency. The Health Maintenance Organization (HMO), moreover, shares the patient's incentives—stay healthy and reduce the cost of care.

Private health agencies cannot, however, be expected to finance the basic research that underlies much practical health information. Many health information products, moreover, are needed by almost everyone and are expensive to produce. Consider the interactive video program on benign prostatic hyperplasia (BPH) described by Wennberg (chapter 6). It may not be economical for each care-providing organization to develop its own.

In addition to the issue of cost, different institutions have different incentives regarding the presentation of information to prospective consumers of health care services. Returning to the BPH example, fee-for-service providers may not wish to support a program that has been demonstrated to reduce the demand for surgery. Similarly, as Wennberg points out, a pharmaceutical company that sells drug treatments for BPH is not a credible source of information about BPH treatment options. The HMO, with an incentive to keep costs down, may be tempted to skew the information in favor of low-cost options such as watchful waiting and to downplay or even omit reference to more expensive treatment alternatives.

Information that reduces the frequency of visits to the doctor, or that

---

[7]One should also note the delicate issue of liability. Should an automated system suggest to a patient that a visit to the doctor is not necessary, when in fact urgent care is required, the information provider may be liable for any damages that arise due to neglect. Zallen acknowledges that the system operated by Harvard Community Health Plan was purposely designed to be very conservative, presumably for this reason.

leads to a choice of a less expensive treatment, has a short-term payoff for the managed care provider. The economic incentives of long-term health promotion are not so clear. Information that promotes healthy behavior and reduces the incidence of disease many years in the future has a clear social benefit, but this benefit may not be captured by an individual's current health care provider. In the example of teenage obesity, which may lead to middle- aged heart disease, the provider serving the teenager is not likely to be the same provider who will cover the cost of open heart surgery in 30 years. Thus whereas the expensive treatment may have been avoidable through effective health education at an early age, the teenager's provider would not appear to have an incentive to provide such education. Information with longer term benefits would thus appear to require public support.

Another issue is the comprehensibility of the information itself. Suppose that all of the health information an individual could ever need is easily accessible, but that all of it is written for a college reading level. Many who have access to this information would then not be served by it. As a number of chapters in this book have suggested, the new media create opportunities to present information in ways that are most appropriate to the learning style of the individual, including information in video format. However, despite the ease of manipulation afforded by digital media, each new adaptation of information to a particular audience will carry with it a marginal cost. At some point, this marginal cost will exceed the marginal benefit associated with making the information accessible to a particular group. Without social support, this group, and those groups whose associated marginal benefits are even smaller, may not be able to take full advantage of the health information made available on the information highway.

The Role of the Public Sector

The preceding discussion suggests a number of areas related to health information where the private sector may not by itself serve the needs of the public. Two potential government roles stand out: ensuring the credibility of health information and producing the packages of health information that the private sector has little incentive to produce.

Different institutions in a reformed health care system will likely have different biases when it comes to the presentation of information that affects consumers' choices. These biases can result in a consumer being faced with conflicting information from different sources, with little clue as to the accuracy of what they read, see, or hear. The government may have a role in helping the consumer determine what and whom to believe.

One approach to the credibility problem is to have the government produce health information and disseminate it directly or through intermediaries. To the extent that the government is a credible source, the consumer's concern over what to believe is alleviated. An alternative option is to have the government bless certain information produced by the private sector. Just as the department of agriculture and the food and drug administration currently assure consumers of the quality of the meats they eat or of the safety of the medications they take, the government could, in effect, provide a seal of approval for health information. Rather than a mandatory process, the government could offer private health information producers the opportunity to have their products reviewed and approved by the government. The incentive for the information producer would depend on the value (in terms of enhanced credibility and possible protection from liability) that the approval may bring.

The second major public role is in the production of information that can have social benefits but that is not likely to be produced by the private sector. As discussed above in the example of teenage obesity and heart disease, some health information may have payoffs that are too distant to elicit private sector investment. The net social benefit may be positive, but the net benefit to any one firm is not. A similar case is information targeted at a small audience, where the marginal cost of production exceeds the marginal benefit to the firm of reaching that audience. Government production of such information may be justified if it can reach a sufficiently large audience to offset the cost of production. For example, a particular HMO in the Minneapolis area may not be able to justify a multimedia tutorial on prenatal care aimed at a handful of its patients from the Hmong community. However, the federal government may be able to justify the production of the tutorial if it can be used in Hmong communities throughout the country. In either of these cases, the test for the government seems fairly clear. If there is a net social benefit associated with the production of the information, and the information is not being produced by the private sector, then the government should produce it.

Nearly all proposals for health care reform rely largely on market mechanisms to bring rationality to the nation's system of care. Market theory, in which rational consumer choices reward efficient and punish inefficient firms, presupposes that consumers have perfect information about the goods and services they purchase. The Clinton administration, in its health care reform proposal, sought to improve the rationality of consumer choices by providing more information on the performances of different health plans, doctors, and hospitals (White House Domestic Policy Council, 1993). Consumers would presumably shun providers that perform poorly and charge high fees in favor of those that offer better quality at lower rates if the facts were accessible to them.

Other decisions about their medical care besides shopping for quality health providers relate to such matters as whether to undergo surgery, or whether to avail one's self of prenatal care. These decisions also are not always rational in the economic sense. Again, economically rational decisions are based on perfect information, and, as we know, health information is not always perfect. If the market does not provide adequate information, then it is the responsibility of the public sector to reduce the gaps in information that lead to irrational choices. To do so, the government needs to assure that its citizens have access to the information that they need and that that information does not mislead.

CONCLUSION

This chapter has challenged the vision developed in the rest of the book in two respects, questioning whether, without stronger public leadership: (a) the National Information Infrastructure will come into being with the power of switched video, universal reach and openness needed for health purposes, and whether (b) the needed health information products to be accessed over that media will be developed.

A central conclusion is that the free market alone is not likely to do these jobs. Private incentives do not reflect the public externality values that accrue to public health. The interests of private firms, moreover, may even work against providing openness in the infrastructure or developing information products with long-range benefits of value to a small and widespread constituency.

To assure that the new media adequately serve health requires leadership beyond that of the health community. Infrastructure issues must be faced at the top levels of different branches of the federal government: *executive*—especially the president's office of science and technology policy, the office of management and budget, the department of commerce national telecommunications and information administration, and the interagency coordinating committees that have been created in the federal government for information infrastructure issues; *regulatory*—the Federal Communications Commission; and *legislative,* particularly the congressional committees dealing with telecommunications issues.

State governments also are seriously involved, as they set the goals of state information networks, and, through state utility commissions, define universal service and interconnection rights and approve network investments. Municipalities, as they grant franchises to television and telephone companies, have an opportunity to require universal service and openness of the information infrastructure.

Although those of us concerned with health do not have such responsibilities for government policy, we can help improve the chances of getting the infrastructure needed for health purposes by specifying what health needs are and what powers and reach of the infrastructure they require.

Unfortunately, both private and public leadership in media development tend to define what is needed based only on demonstrated customer demand. This has meant that the design of the information infrastructure is being derived largely from present and projected payments for telephone service and for video entertainment and home-shopping (Johnson & Reed, 1990).

Such a view overlooks public externalities, it is like deciding whether a town needs a second high school by asking students how much they would pay to enter its doors. But whereas estimating future demand based only on past experience is limiting, it is understandable. After all, telecommunications experts cannot be expected to imagine health uses of the new media of which they know nothing, uses that may never even have been tried out. That leaves a large role for the health community, first to be imaginative, then to experiment, and finally to inform those making policy about the new uses that new media could help provide.

In that sense, the two subjects of this chapter should be reversed. We should first foster development of information programs that could use the media to advance health. Yes, even before that media is in place. Right now every health activity should be setting aside some time, and making some investment to explore how its health purpose could be pursued given a universal broadband infrastructure that reaches every home.

The result of such forward-looking activities by the health community would greatly improve our ability to perform our second task: to describe to the political leadership the needed power and reach of the media, for we could cite many specific potential uses. Some of the activities described in this book are steps in exactly the right direction. But the development of health uses of information must be taken further, and taken up by others. We need more pioneers, more visions of the future, or we won't persuade our leaders to give us the information superhighway required to improve the nation's health.

REFERENCES

Computer Systems Policy Project. (1993). *Perspectives on the national information infrastructure: CSPP's vision and recommendations for action.* Washington, DC.

Electronic Frontier Foundation. (1993). Open platform campaign: Public policy for the information age. Washington, DC.

Fisher, F. D. (1994, April). *Identifying the potholes in the information superhighway: A public interest perspective. Telecommunications, 28*(4), 23.

Gore, A. (1994, January 11). Remarks delivered at UCLA. White House, Washington, DC.

Hadden, S. G. for the Alliance for Public Technology. (1994). Extending universal service through the NII. In L. I. Leidig (Ed.), *20/20 Vision: The development of a national information infrastructure* (pp. 47–54). Washington, DC: U.S. Department of Commerce, National Telecommunications and Information Administration.

Information Infrastructure Task Force (1993). *The national information infrastructure: Agenda for action.* Washington, DC: U.S. Department of Commerce, National Telecommunications and Information Administration.

Johnson, L. L., & Reed, D. P. (1990). *Residential broadband services by telephone companies? Technology, economics and public policy.* Santa Monica, CA: RAND.

Ministry of Supply and Services, Canada. (1992). *Convergence, competition and cooperation: Policy and regulation affecting local telephone and cable networks.* Report of the Co-chairs of the Local Networks Convergence Committee, Ottawa, Ontario.

National Cable Television Association, Inc. (1993). Brief in the U.S. Supreme Court on appeal from the U.S. District Court (DC). In *Turner Broadcasting System, Inc. v. F.C.C.* (p. 36). Washington, DC.

National Research Council, Computer Science and Telecommunications Board. (1994). *Realizing the information future: The Internet and beyond.* Washington, DC: National Academy Press.

New York Telecommunications Exchange. (1993, December). *Connecting to the future: Greater access, services, and competition in telecommunications* (Report). New York: Office of Economic Development.

Office of Technology Assessment. (1991). New ways of configuring rural networks. In *Rural America at the crossroads: Networking for the future* (pp. 82–87). L. Garcia, project director. Washington, DC: U.S. Government Printing Office.

Pingree, S., Hawkins, R. P., Gustafson, D. H., Boberg, E. W., Bricker, E., Wise, M., & Tillotson, T. (1993). Will HIV-positive people use an interactive computer system for information and support? A study of CHESS in two communities. In *IEEE Proceedings of the 17th Annual Symposium on Computer Applications in Medical Care,* Washington, DC.

Smith, R. W. (1993, October 27). Statement of Raymond W. Smith, Bell Atlantic Corp., on merger with Tele-Communications Inc. before Subcommittee on Antitrust, Monopolies & Business Rights of the Committee on the Judiciary, U.S. Senate. Washington, DC.

Time Warner, Inc. (1993, March 17). Press release on "full service network" for Orlando, Florida. New York.

U.S. Department of Health and Human Services, Public Health Service. (1991). *Healthy people 2000: National health promotion and disease prevention objectives.* Full report with commentary (DHHS Pub. No. (PHS)91-50212). Washington, DC: U.S. Government Printing Office.

White House Domestic Policy Council (1993). *Health security: The President's report to the American people.* Washington, DC.

GLOSSARY

The New Media: Annotated Glossary

Julia A. Marsh
Lawrence K. Vanston
Technology Futures, Inc.

ATM (Asynchronous Transfer Mode)—see Switching

Bandwidth Bandwidth refers to the rate at which information is transmitted over a communications medium. Higher bandwidth equates to more information per time interval. It is normally measured in bits per second (b/s)—that is, the number of 0s and 1s that can be transmitted in a second. These speeds are usually reported in thousands (kilobits per second or Kb/s), millions (megabits per second or Mb/s) , or billions (gigabits per second or Gb/s). Generally, the higher the bandwidth of a channel, the more it costs and the harder it is to obtain. Ranges of bandwidth are categorized in Table 1.

The bandwidth required for interactive multimedia varies according to a number of factors. Breaking up multimedia into its components is a good way to understand the bandwidth requirements. Narrowband provides excellent quality for voice and audio communications, and is also very suitable for data and text. However, when images need to be sent, narrowband becomes inadequate.

TABLE 1
Categories for Digital Data Rates

Category	Range of Speeds	Typical Speed	Terminology
Narrowband	56 Kb/s to 128 Kb/s	56 or 64 Kb/s	DS0
Wideband	256 Kb/s to 10 Mb/s	1.5 Mb/s	DS1 or T1
Broadband	10 Mb/s and above	45 Mb/s	DS3

The line between acceptable and unacceptable quality is subjective and depends greatly upon the application. Generally, narrowband will provide crude video that might be acceptable for desktop conferencing or small video windows on a computer screen. Wideband is typically used for videoconferencing today because, for businesses, it provides a good tradeoff between quality and cost. Wideband is also suitable for the type of video required for desktop multimedia. However, only the upper reaches of the wideband range can provide full-motion broadcast quality television and HDTV will certainly require broadband. In the long run, broadband with its higher video quality, will become more desirable, especially when the costs come down. For high-end applications, such as some types of medical imaging, broadband is an immediate requirement.

Carriers   In the telecommunications industry, carriers typically refer to the companies providing local and long distance telephone service because their networks carry the signals from point-to-point. AT&T, MCI, and Sprint are the three best-known long distance carriers. The largest local exchange carriers include GTE, BellSouth, Pacific Bell, Bell Atlantic, NYNEX, Ameritech, Southwestern Bell, and U.S. West.

Communications Act of 1934   This act gave the Federal Communications Commission (FCC) regulatory authority over the telephone industry. Under this act, each local operating telephone company is guaranteed an area of operation without competition and assured a certain maximum rate of return; local operating companies must connect with anyone within their operating area requesting telephone service; and charges for local telephone service are subject to government approval through tariffs filed with the FCC or state regulatory bodies.

Confidentiality—see Data Security

Copper Wires   These are the two wires of a circuit, in particular the wire pairs in telephone cables. A cable consists of a number of pairs of copper wire conductors assembled together in a compact form and bound together with a strong, flexible, waterproof sheath.

Copyright   Copyright laws provide protection to the authors of "original works of authorship" including literary, dramatic, musical, artistic, and certain other intellectual works. This protection is available to both published and unpublished works (U.S. Copyright Office, 1992).

Data Security   Data security involves the preservation of the confidentiality, reliability, and protection of personal information. This is particularly important for sensitive patient information.

Digital   Digital signals constitute a discontinuous stream of on/off (1s and 0s) pulses, as opposed to analog, which transmits continuous wave forms. Digital transmission speeds are usually measured in bits per second (b/s). Digital signals can be carried over a number of different telecommunications media including copper wires, fiber optics, coaxial cable, microwave, and satellite.

Fiber Optics   Fiber optic systems transmit information through pulses of light energy rather than through electrical energy. The light energy is transmitted over an extremely clear, thin glass optic fiber. Each of these fibers can carry hundreds of thousands of voice circuits. The relative advantages of fiber optics include greater capacity, reduced repeater requirements over distances, immunity to electrical interference, lower maintenance costs, and greater reliability.

Industry Convergence   The integration of these information types—textual, audio, and visual—forces the technologies and markets of three multibillion dollar industries to converge, namely computer, communications, and entertainment. The computer industry has the expertise to develop the multimedia hardware and software. The communications industry—telephone, cable, and satellite—have the expertise for delivering information from point-to-point or point-to-multipoint. The entertainment industry—television, movie, publishing—owns much of the information content. They have the expertise and resources to develop new multimedia-based content, or integrate existing content. A fourth key industry impacted by these technologies is consumer electronics. Companies like Nintendo and Sega already have an installed base in 30 to 40% of all U.S. households. They also have the resources to bring educational information as well as entertainment into the home.

Intellectual Property Rights   Rights that protect your intellectual property under the copyright, trademark, and trade secret laws.

Interactive Multimedia   A key capability of the computer is the element of interactivity. By adding a peripheral such as a keyboard or a joystick, one can begin to interact with computer-mediated information. This increased involvement graduates one from simply a **passive viewer** to an **active user** of information. "Interactive multimedia enables a user to respond to what is presented and effect what is presented next in a combination of full-motion video, still video, graphics, animation, text, and sound or at least a selection of these. The primary key to success with an interactive multimedia offering is to master the art/science of *interactivity* in such a way that it will be perceived as interesting and beneficial to a large number and wide variety of users" ("What is Hypermedia," 1990).

Multimedia becomes *hypermedia* when you make logical connections ("links") between these information types ("nodes") to create an interactive database. Hypermedia, in its most general form, is an electronic database that connects text, graphics, audio, and video segments or nodes of information using associative links. It is an extension of a form of electronic cross-referencing called hypertext. These links allow us to navigate through a hypermedia database in a nonlinear fashion selecting only those nodes of information that may be of interest. It stimulates us to explore, create, and manipulate information in a self-guided, self-paced environment.

Interconnectivity   Demand for two-way interactivity and resource sharing has driven the need for interconnection of hardware and networks. Connecting computers and peripherals, such as printers and modems, over a short distance is called a local area network. Connecting one network with another network is called "internetworking." As these industries and technologies converge, it is imperative that this "network of networks" be able to communicate and interface seamlessly. Just as we are able to call any other telephone or fax machine, these new computer-TV-telephone combination machines must be just as flexible and as easy to use. The value in transparent interconnection lies in the user's ability to connect to "anyone, anywhere" without regard to hardware or software.

Interoperability   Demand to interconnect hardware and networks is driving the need for interoperability. This necessitates a common language or protocol that enables different hardware or networks to communicate with one another. A classic example of interoperability today is the fax machine. We can send faxes to anyone without consideration of hardware or software on either end.

ISDN (Integrated Services Digital Network)   ISDN (Integrated Services Digital Network) is a planned hierarchy of digital switching and transmission systems synchronized so that all digital elements speak the same "language" at the same speed. The ISDN will provide voice, data, and video in a unified manner.

The *narrowband ISDN* represents a set of international standards that specifies how subscribers can access a public network that supports digital, switched telecommunications services. The basic rate ISDN user-to-network interface provides two 64 Kb/s channels (called B channels) and one 16 Kb/s signaling channel (the D channel). This basic rate interface could simultaneously support voice communications on one B channel, image communications on the other, and call progress or other information on the D channel. Major long distance carriers are offering switched, narrowband services on their networks, although internetworking and standardization are problems. Even the availability of ISDN, however, will not solve the communications problems for networked interactive multimedia. First, it is narrowband, which provides only marginal multimedia communications. Second, it is circuit switched, which is not particularly good for applications that are data oriented.

Broadband ISDN or *BISDN* is the likely solution for a universally available public switched broadband network. BISDN services provide customers with transmission rates ranging from 45 Mb/s to 622 Mb/s. It will serve the communications needs of a mixture of voice, data, image, and video applications. However, it may be a decade before BISDN becomes widely and economically available. In the long run, a combination of LANs, private WANs, and public BISDN will provide an almost transparent communications infrastructure for interactive multimedia that will be both economical, universal, and completely standardized.

Local Area Networks (LANs)   A LAN is a telecommunications network permitting the interconnection and intercommunication of a group of computers and terminals. Most LANs are privately owned and operated within a limited geographic area: a physician's office, within a hospital, or between a hospital and a satellite clinic. They are designed to specifically provide high-speed, high-bandwidth data transmission. A LAN provides a number of different services for the computers and users it links together. These services usually include file sharing and electronic mail. Some networks offer just a few of these services, and others offer many more.

Because LANs are designed for data communications, as opposed to multimedia, they have limited bandwidth. Although progress is being made in adapting LANs to multimedia, it is likely that multimedia will eventually lead to a new generation of local area networks. When local area networks are interconnected within a metropolitan area they become *Metropolitan Area Networks* (MANs).

When an organization is geographically dispersed, it needs a **wide area network (WAN)** to connect its LANs. This requires the involvement of telecommunications carriers to provide the connections between the locations. Today, these connections are almost always leased private-lines that provide a permanent digital "pipe" between a pair of locations. Since these pipes may carry the bulk of communications between the two locations they are often high-speed, usually wideband, and, thus, expensive. Because today's wideband WANs are shared among many users, they face the same problem that LANs do when it comes to providing multimedia communications—not enough bandwidth.

Metropolitan Area Networks (MANs)—see Local Area Networks (LANs)

Multimedia   Even with all the publicity and rhetoric about the power
and potential of multimedia, it is difficult to get a clear, concise definition of
exactly what it means. Webster's defines *multi* as more than two and *media*
as all the means of communication, such as newspapers, radio, and TV, that
provide the public with information and entertainment. Therefore, multimedia could be defined as the combination of various kinds of media for the
same purpose; for example, motion pictures with stereo sound to produce
the maximum effect on an audience.

The product of a computer-based communication system is typically
information. Consequently, to create a multimedia environment at least two
or more of the basic information types must be integrated. The basic information types include: textual (text and numbers), audio (speech and music),
and visual (photographic images, video, graphics, and animation).

By combining audio and/or video information with textual information
we can make it multidimensional and multisensory. This enables us to
communicate in a manner more closely approximating the sights and sounds
of the world around us.

But this alone is not enough. After all, we have seen multimedia information on our TV and at the movies for years. It is the advent of the personal
computer that makes the integration of these information types so exciting
and intriguing. When the visual power of the television is married with the
computer, the possibilities become virtually limitless. We can begin to create
information that has greater effectiveness, impact, and persuasiveness.

National Information Infrastructure (NII)   The NII is a Clinton administration initiative to support the private sector construction and maintenance of a "seamless web of communications networks, computers, databases, and consumer electronics that will put vast amounts of information at
users' fingertips." The effort is guided by nine principles and objectives: (a)
promote private sector investment; (b) extend the 'universal service' concept to ensure that information resources are available to all at affordable
prices; (c) act as catalyst to promote technological innovation and new
applications; (d) promote seamless, interactive, user-driven operation of the
NII; (e) ensure information security and network reliability; (f) improve
management of the radio frequency spectrum; (g) protect intellectual property rights; (h) coordinate with other levels of government and with other
nations; (i) provide access to government information and improve government procurement.

Two groups have been established to support this initiative: (a) The
Information Infrastructure Task Force (IITF), made up of high-level representatives of the federal agencies that play a major role in the development
and application of information technologies; and (b) The U.S. Advisory
Council on the National Information Infrastructure, made up of senior-level

representatives of the private sector, including business, labor, academia, public interest groups, and state and local governments, to facilitate private sector input to the IITF (Brown, 1993).

Networked Interactive Multimedia   A growing need to exchange information and share computing resources has resulted in networked computing. Networks are essential to two-way interactive multimedia communications. Through networking, time, geographic, and distance constraints are reduced. With its mix of video, audio, graphics, and text, networked interactive multimedia needs robust and flexible communications. In all practicality, this means high-speed, switched, two-way digital communications. This will ultimately mean a universally-available public switched digital network which facilitates "anywhere, anytime" communications with anyone.

Personal Communications Services (PCS)   Personal communications services refer to wireless communications with a high level of portability, for example, cordless and cellular phones. It envisions: system capacity and coverage to eventually serve everyone, a price low enough to encourage mass usage, a personal number that allows a person to be reached no matter where the person is located, quality and reliability at least commensurate with today's cellular service, intelligent features that allow the user control over who can reach him—and when and where—along with integrated voice mail, data, and fax capabilities.

Privacy   In the U.S., a person's right to personal privacy is protected by the constitution. However, there is little or no protection of personal data, except in such specialized areas as credit reporting. As networks interconnect and information becomes more universally accessible, the security, reliability, and privacy of sensitive information comes into question.

Private Networks   Private networks are typically used by companies that are geographically dispersed with high volume communications needs, such as between a metropolitan hospital and satellite clinics. The line is dedicated for the sole use of company communications. Usage is typically billed on a flat monthly rate rather than by time-based usage. Historically, they offered a cheap alternative to the public-switched network, but that is beginning to change.

There are three problems with private networks. First, since the organization has to pay for the full-time use of an expensive communications channel, they are not cost-effective for locations that don't have much traffic. Second, there is no practical way to communicate at high speed with anyone that is not connected to the network. Although most communications may be internal, many of the most important ones may be external to the organization. Third, they are not a practical solution for small organizations or individuals, including those working at home.

Public Networks   With a public network, anyone can access the network and communicate with anyone else on the network whenever they want and pay for communications only when they use them. A universally-available network enables "anywhere, anytime" communications with anyone.

Regulatory Agencies   Because the local exchange carrier constitutes a natural monopoly over local telephone services, it is regulated at both the state and federal levels. At the state level, the local telephone company is regulated by a commission. At the federal level, it is regulated by the Federal Communications Commission (FCC).

Reliability—see Data Security

Satellite   Launched into orbit to function as electronic relay stations, satellites are based on radio frequency (RF) signals found in the electromagnetic spectrum. Orbiting at an altitude of 22,300 miles above the earth's surface, they appear stationary when observed from the ground. Initially, satellites primarily carried telephone calls, which require relatively low-powered transmissions. As the capacity of satellites to carry more signals grew, television signals—which require more power and bandwidth—were relayed by satellite. By the 1980s, high-volume data transmissions became possible through satellite communications (Grant & Wilkinson, 1993–1994)

Scalability   Scalability refers to the ease with which a piece of hardware or software can be upgraded cost-effectively. This is a critical component of today's rapidly evolving technologies. For example, we would say a networked medical imaging system is scalable if it can be easily expanded to serve more physicians or patients than originally planned or can be upgraded with newer versions of software and hardware without replacing the entire system.

Stand-Alone Information Appliances   The delivery of multimedia information can take many forms. The first generation of multimedia products has typically been delivered via stand-alone hardware, such as VCR, laser disc, and CD-ROM players. This first generation of multimedia information appliances is fundamental to the development and commercialization of multimedia-based software, but there are drawbacks. Information stored on laserdiscs, VHS tape, or CD-ROM cannot be updated; therefore, it becomes time-dated and static. A stand-alone unit may be sufficient for a tutorial on breast self-examination, but totally inadequate when demand requires access to dynamic, real-time information. One example of this would be a collaborative consultation between a specialist at the regional teaching hospital and a physician located in an outlying rural community.

Standards Standards are promises made by software and hardware manufacturers to content developers that tomorrow's machines and programs will still be able to use the information—textual, audio, and visual—created on today's machines with today's programs. Multimedia computer standards, which enable cross-platform compatibility, will help encourage system and application software system development. Standards will also provide hardware manufacturers with specific guidance on what to build, thereby providing developers with a clear framework for the design and development of applications.

Switching *Switched* means that you can choose to communicate with different parties at different times. A nonswitched channel, on the other hand, provides continuous communications between two points.

There are two basic types of switching used in telecommunications today: *circuit switching* and *packet switching*. With circuit switching, a continuous communications path, a circuit, is set up for the duration of a call. With packet switching, information is assembled in packets and sent over the network only when there is something to send. Circuit switching is most appropriate for communications such as telephone conversations or video conferences that are more or less continuous for the duration of the call. Packet switching is better for bursty communications that have alternating idle and busy periods. Data communications are the typical application for packet switching. For example, a remote user of a database of information may be on-line for hours, but actually retrieve large amounts of data sporadically. Packet switching can, however, introduce small delays when sending a message. This is not usually a problem with data transmissions, but is disastrous for voice and video communications.

Interactive multimedia poses some important issues for switching. Some of the elements, such as voice and videoconferencing, need circuit switching, whereas other elements such as text, image, and video clip retrieval are better suited for packet switching. Which type is best will depend on the particular multimedia application. Applications that are oriented mostly toward information retrieval will be best suited for packet networks. On the other hand, applications that depend heavily on interactive communications between people, using voice and video, will be better suited for circuit switching.

A new type of switching called **Asynchronous Transfer Mode (ATM)** will ultimately bridge the gap between packet switching and circuit switching. ATM uses packets called cells, whose format and length are chosen to be efficient for all types of communications. ATM switches can switch cells so fast that there is no perceptible delay. Thus, ATM switching can create the illusion of circuit switching for those applications that need continuous communications for voice or video. They will enable the same networks to be used for all types of applications. This not only will facilitate more efficient networks, but will also greatly simplify the design of multimedia applications.

Wide Area Networks (WANs)—see Local Area Networks (LANs)

REFERENCES

Brown, R. H., Irving, L., Prabhakar, A., & Katzen, S. (1993, September). *The national information infrastructure: Agenda for action.* Washington, DC: U.S. Department of Commerce.

Grant & Wilkinson (Eds.). 1993–1994. Satellites. *Communication Technology Update.*

U.S. Copyright Office (1992). *Copyright basics* (Publication No. 312-433/60, 021). Library of Congress. Washington, DC: U.S. Government Printing Office.

What is hypermedia? (1990, April 30). *The Seelinger Letter, 2.*

Author Index

Subject Index

helping people learn, 193
professional education, 197
self-care, 57
NII, *see* National Information Infrastructure
NLM, *see* National Library of Medicine
Nodes, interactive multimedia, 234
Nonhealth motives
demand management, 52
self-care relation, 53
NPR, *see* National Performance Review
NSF, *see* National Science Foundation

O

Objectivity, shared decision making
programs, 121–123
Options
cost and decision making, 50
treatment and shared decision making,
115–120, 121, 125
Orphan risk behaviors, private vs. public
obligations, 139, *see also* Behavior
Outcomes research
health information for public, 134
potential, 109
quality of health care, 7–8
shared decision making, 113–118, 125
Overload, *see* Information overload
Overspecialization, health care systems, 9

P

Packet switching, about, 239
Paradigmatic illustrations, learning in
simulated environments, 178
Password, security in Harvard Community
Health Plan, 35
Pathology, personal and multimedia, 167
Patient contact, triage system of Harvard
Community Health Plan, 25–30
Patient focus groups, shared decision
making, 120
Patient preference
decision making, 123
shared, 110–111
subjectivity, 112, 113, 122
demand management, 49–52
Patient service, member-centered care
systems, 38
Patients, empowering and decision making,
113–118

Payout, Texas Telemedicine Project, 79, 80
PCS, *see* Personal Communications Services
Peers
collaboration, learning, 166
communication through bulletin boards,
215
education and interactive multimedia,
147, 150
support, HealthWorld, 177
Perceived need
demand management, 48–49
self-care relation, 53
Personal Communications Services (PCS),
about, 237
Physical activity, multimedia product by
ABC News Interactive, 153
Physician preferences, medical decision
making, 51
Pioneer Communications, training in
multimedia using ABC News
Interactive programs, 155–159
Pipe, Wide Area Networks, 235
Positive/negative findings, triage system of
Harvard Community Health Plan,
27, 29
Practice variations, patterns, 109–110
Preadmission certification, hospitalization
reduction, 45
Preference trials, shared decision making, 125
Prescription, automated and Harvard
Community Health Plan, 31
Prevention
cost-reduction strategy, 46
exploitation of telemedicine for, 86
infectious diseases and unintentional
injury, 47
mortality and risk factors, 128–130
Privacy, issue in quality of health care
systems, 8, 237
Private networks, about, 237–238, *see also*
Networks
Private sector, *see also* Federal government;
Public sector
content review and assistance by federal
government, 139
funding of National Information
Infrastructure, 135, 136
health information provision, 138, 221–
223
payoff and investment in health information, 224

Printed in the USA/Agawam, MA
August 30, 2013

579511.002